MATHEMATICS RESEARCH DEVELOPMENTS

RECENT STUDIES IN DIFFERENTIAL EQUATIONS

MATHEMATICS RESEARCH DEVELOPMENTS

Additional books and e-books in this series can be found on Nova's website under the Series tab.

MATHEMATICS RESEARCH DEVELOPMENTS

RECENT STUDIES IN DIFFERENTIAL EQUATIONS

HENRY FORSTER
EDITOR

Copyright © 2020 by Nova Science Publishers, Inc.

All rights reserved. No part of this book may be reproduced, stored in a retrieval system or transmitted in any form or by any means: electronic, electrostatic, magnetic, tape, mechanical photocopying, recording or otherwise without the written permission of the Publisher.

We have partnered with Copyright Clearance Center to make it easy for you to obtain permissions to reuse content from this publication. Simply navigate to this publication's page on Nova's website and locate the "Get Permission" button below the title description. This button is linked directly to the title's permission page on copyright.com. Alternatively, you can visit copyright.com and search by title, ISBN, or ISSN.

For further questions about using the service on copyright.com, please contact:
Copyright Clearance Center
Phone: +1-(978) 750-8400   Fax: +1-(978) 750-4470   E-mail: info@copyright.com.

NOTICE TO THE READER

The Publisher has taken reasonable care in the preparation of this book, but makes no expressed or implied warranty of any kind and assumes no responsibility for any errors or omissions. No liability is assumed for incidental or consequential damages in connection with or arising out of information contained in this book. The Publisher shall not be liable for any special, consequential, or exemplary damages resulting, in whole or in part, from the readers' use of, or reliance upon, this material. Any parts of this book based on government reports are so indicated and copyright is claimed for those parts to the extent applicable to compilations of such works.

Independent verification should be sought for any data, advice or recommendations contained in this book. In addition, no responsibility is assumed by the Publisher for any injury and/or damage to persons or property arising from any methods, products, instructions, ideas or otherwise contained in this publication.

This publication is designed to provide accurate and authoritative information with regard to the subject matter covered herein. It is sold with the clear understanding that the Publisher is not engaged in rendering legal or any other professional services. If legal or any other expert assistance is required, the services of a competent person should be sought. FROM A DECLARATION OF PARTICIPANTS JOINTLY ADOPTED BY A COMMITTEE OF THE AMERICAN BAR ASSOCIATION AND A COMMITTEE OF PUBLISHERS.

Additional color graphics may be available in the e-book version of this book.

Library of Congress Cataloging-in-Publication Data

Names: Forster, Henry (Editor of Nova), editor.
Title: Recent studies in differential equations / Henry Forster (editor).
Identifiers: LCCN 2020032609 (print) | LCCN 2020032610 (ebook) | ISBN
  9781536183894 (paperback) | ISBN 9781536184297 (adobe pdf)
Subjects: LCSH: Differential equations.
Classification: LCC QA371.3 .R47 2020 (print) | LCC QA371.3 (ebook) | DDC 515/.35--dc23
LC record available at https://lccn.loc.gov/2020032609
LC ebook record available at https://lccn.loc.gov/2020032610

Published by Nova Science Publishers, Inc. † New York

CONTENTS

Preface		vii
Chapter 1	Almost Automorphic and Asymptotically Almost Automorphic Type Functions in Lebesgue Spaces with Variable Exponents $L^{p(x)}$ *Toka Diagana and Marko Kostić*	1
Chapter 2	Existence and Regularity of Solutions for Some Nonlinear Second Order Differential Equation in Banach Spaces *Issa Zabsonré and Napo Micailou*	29
Chapter 3	Oscillation Results for Nonlinear Neutral Impulsive Differential Equations *Shyam Sundar Santra*	49
Chapter 4	First-Order Forced Functional Differential Equations *Shyam Sundar Santra*	61
Chapter 5	Oscillation Criteria for Neutral Difference Equations *Shyam Sundar Santra, Debasish Majumder, Rupak Bhattacharjee and Tanusri Ghosh*	85
Chapter 6	PDEs Satisfied by the Density Function of Stochastic Integrals *Julia Calatayud, Juan Carlos Cortés and Marc Jornet*	107
Index		121

PREFACE

This compilation introduces and studies the class of (asymptotically) Stepanov almost automorphic functions with variable exponents, presenting a few relevant applications of abstract Volterra integro-differential inclusions in Banach spaces.

The authors study the existence and regularity of solutions for some nonlinear second order differential equations, showing the existence of mild solutions and giving sufficient conditions ensuring the existence of strict solutions.

Sufficient conditions for the oscillation of solutions of neutral impulsive differential equations are also presented.

In the penultimate study, the oscillatory behaviour of the solutions of a class of nonlinear first-order neutral differential equations with several delays of one form are studied.

In addition, some sufficient conditions for the oscillation of solutions to the first and second-order neutral delay difference equation are presented.

Chapter 1 introduces and studies the class of (asymptotically) Stepanov almost automorphic functions with variable exponents. Any function belonging to this class needs to be (asymptotically) Stepanov almost automorphic. A few relevant applications to abstract Volterra integro-differential inclusions in Banach spaces are presented.

Chapter 2, studies the existence and regularity of solutions for some nonlinear second order differential equation. The delayed part is assumed to be locally lipschitz. Firstly, the authors show the existence of the mild solutions. Secondly, they give sufficiently conditions ensuring the existence of strict solutions.

Chapter 3 presents sufficient conditions for oscillation of solutions of neutral impulsive differential equations (IDEs)

$$\begin{cases} \frac{d}{dt}[x(t) - r(t)x(t-\tau_1)] + q(t)H(x(t-\sigma_1)) = 0, & t \geq t_0,\ t \neq \tau_k \\ x(\tau_k^+) = J_k(x(\tau_k)), & k \in \mathbb{N} \\ x((\tau_k - \tau_1)^+) = J_k(x(\tau_k - \tau_1)), & k \in \mathbb{N} \end{cases}$$

for $|r(t)| < +\infty$.

In Chapter 4, oscillatory behaviour of the solutions of a class of nonlinear first-order neutral differential equations with several delays of the form

$$\bigl(x(t) + p(t)x(t-\tau)\bigr)' + \sum_{i=1}^{m} q_i(t)H\bigl(x(t-\sigma_i)\bigr) = f(t)$$

are studied. This problem is considered in various ranges of the neutral coefficient p. The main tools are Knaster-Tarski fixed point theorem and Banach's fixed point theorem. Examples are provide to show feasibility and effectiveness of the main results.

Chapter 5 presents some sufficient conditions for the oscillation of solutions to the first and second-order neutral delay difference equation. Also, the authors state some open problems and some examples are presented to show effectiveness of the main results.

In Chapter 6, the authors derive a partial differential equation (PDE) for the density function of stochastic processes of the form $X(t) = \int_{t_0}^{t} Y(s)\,\mathrm{d}s$, $t \geq t_0$, where Y is any stochastic process with values in $\mathbb{R}$. The resulting PDE depends on the conditional expectation of $Y(t)|X(t) = x$, which is in general unknown. However, the PDE is applicable in the case of random differential equation problems, which yields Liouville's equation for the density of the solution. It is also applicable to deduce a PDE analog of the random variable transformation method. On the other hand, when Y is Gaussian, such conditional law $Y(t)|X(t) = x$ is known; therefore explicit PDEs can be obtained and tested on important Gaussian processes (Brownian motion, Brownian bridge, white noise, Ornstein-Uhlenbeck and fractional Brownian motion processes). In particular, when $Y(t) = h(t)W(t)$, where h is a deterministic function and W is a Gaussian white noise process, the authors find a PDE for the density function of Itô-type integrals, $X(t) = \int_{t_0}^{t} h(s)\,\mathrm{d}B(s)$, where B is a Brownian motion.

Chapter 1

ALMOST AUTOMORPHIC AND ASYMPTOTICALLY ALMOST AUTOMORPHIC TYPE FUNCTIONS IN LEBESGUE SPACES WITH VARIABLE EXPONENTS $L^{p(x)}$

Toka Diagana[1,*] *and Marko Kostić*[2,†]
[1]Department of Mathematical Sciences,
University of Alabama in Huntsville,
Sparkman Drive, Huntsville, US
[2]Faculty of Technical Sciences,
University of Novi Sad, Novi Sad, Serbia

Abstract

The chapter introduces and studies the class of (asymptotically) Stepanov almost automorphic functions with variable exponents. Any function belonging to this class needs to be (asymptotically) Stepanov almost automorphic. A few relevant applications to abstract Volterra integro-differential inclusions in Banach spaces are presented.

Keywords: Lebesgue spaces with variable exponents, Stepanov almost automorphy with variable exponents, asymptotical Stepanov almost automorphy

---
[*]Corresponding Author's Email: toka.diagana@uah.edu.
[†]Corresponding Author's Email: marco.s@verat.net.

with variable exponents, abstract Volterra integro-differential inclusions, abstract (multi-term) fractional differential equations

AMS Subject Classification: 34C27, 35B15, 46E30

1. INTRODUCTION AND PRELIMINARIES

The main aim of this paper is to continue our recent research of Stepanov $p(x)$-almost periodicity and asymptotical Stepanov $p(x)$-almost periodicity raised in [1], as well as to initiate the study of generalized almost automorphy and generalized asymptotical almost automorphy that intermediate the classical and Stepanov concept. This is accomplished here by examining the notion of Stepanov $p(x)$-almost automorphy and asymptotical Stepanov $p(x)$-almost automorphy. We basically follow the approach obeyed in [1], which enables us to conclude that the introduced classes of functions are translation invariant (Stepanov-like pseudo-almost automorphic functions with variable exponents, which have been analyzed in [2], do not possess this property).

We investigate generalized almost automorphic and generalized asymptotically almost automorphic type functions in Banach spaces using results from the theory of Lebesgue spaces with variable exponents $L^{p(x)}$. For a given measurable function $p : [0, 1] \to [1, \infty]$, we define the notions of an $S^{p(x)}$-almost automorphic function and an asymptotically $S^{p(x)}$-almost automorphic function. In the case that $p(x) \equiv p \geq 1$, the introduced notion is equivalent to the usually considered notion of S^p-almost automorphy and asymptotical S^p-almost automorphy.

The organization and main ideas of this paper are briefly described as follows. In Subsection 1.1, Subsection 1.2 and Subsection 1.3 we collect the basic facts about fractional calculus, multivalued linear operators and Lebesgue spaces with variable exponents $L^{p(x)}$, respectively. Section 2 recapitulates some basic definitions and results about generalized almost periodic and generalized almost automorphic functions. Section 3 starts by recalling the definitions of Stepanov $p(x)$-boundedness and Stepanov $p(x)$-almost periodicity in the sense of [1]. The notion of (asymptotical) Stepanov $p(x)$-almost automorphy is introduced in Definition 3 (Definition 4). It is expected that the notion of (asymptotical) Stepanov $p(x)$-almost automorphy is much more general than that of (asymptotical) Stepanov $p(x)$-almost periodicity, and we explicitly show this in Proposition 3.1 and Proposition 3.2. Several continuous embeddings between various Stepanov $p(x)$-almost automorphic spaces are proved in Theorem 3.3, where it is particularly shown that an $S^{p(x)}$-almost automorphic function has to be Stepanov 1-almost automorphic.

We know that any almost periodic function has to be $S^{p(x)}$-almost periodic for any measurable function $p : [0, 1] \to [1, \infty]$; see e.g., [1]. This is no longer true for almost automorphy, where we perceive some peculiar differences between almost automorphy and compact almost automorphy, proving that the almost automorphy of a function $f : \mathbb{R} \to X$ implies its $S^{p(x)}$-almost automorphy only if we impose the validity of some additional conditions (see Proposition 3.4); all these statements have natural reformulations for asymptotical $S^{p(x)}$-almost automorphy.

In Section 4 we introduce (asymptotically) Stepanov $p(x)$-almost automorphic functions depending on two parameters and formulate related composition principles, thus slightly extending the results obtained in [3], [4] and [5]. Using these composition principles, it is very technical to reword several known results concerning semilinear analogues of the inclusions (10)-(11) and (DFP)$_{f,\gamma}$ considered below (see e.g. [6, Theorem 4-Theorem 8; Theorem 10] for more details). Therefore, we will not consider semilinear Cauchy inclusions in this paper.

Concerning applications, our main results are given in Section 5 where we analyze the invariance of generalized (asymptotical) almost automorphy in Lebesgue spaces with variable exponents $L^{p(x)}$ under the actions of convolution products (see Proposition 5.1 and Proposition 5.2). Although this strengthens some of our previous results, we feel duty bound to mention the difficulty of applying Proposition 5.1 in the case that $p(x)$ is not a constant function. This is no longer the case with the assertion of Proposition 5.2 where the use of ergodic Stepanov components with variable exponents plays a crucial role (see also Example 2 below). We also propose several open problems, questions, illustrative examples and applications of our abstract results.

The standard notation is used throughout the paper. We assume that $(X, \|\cdot\|)$ is a complex Banach space. If Y is also such a space, then we denote by $L(X, Y)$ the space of all continuous linear mappings from X into Y; $L(X) \equiv L(X, X)$. Assuming A is a closed linear operator acting on X, then the domain, kernel space and range of A will be denoted by $D(A)$, $N(A)$ and $R(A)$, respectively.

Let $I = \mathbb{R}$ or $I = [0, \infty)$. By $C_b(I : X)$ we denote the Banach space consisting of all bounded continuous functions $I \mapsto X$, equipped with the sup-norm. The Gamma function is denoted by $\Gamma(\cdot)$ and the principal branch is always used to take the powers; the convolution like mapping $*$ is given by $f * g(t) := \int_0^t f(t-s)g(s)\,ds$. Set $g_\zeta(t) := t^{\zeta-1}/\Gamma(\zeta)$, $\zeta > 0$. For any $s \in \mathbb{R}$, we define $\lfloor s \rfloor := \sup\{l \in \mathbb{Z} : s \geq l\}$ and $\lceil s \rceil := \inf\{l \in \mathbb{Z} : s \leq l\}$.

1.1. Fractional Calculus

The first conference on fractional calculus and fractional differential equations was held in New Haven (1974). Fractional calculus has since gained increasing attention due to its wide application in various scientific fields, such as mathematical physics, engineering, biology, aerodynamics, chemistry, economics etc. Comprehensive information about fractional calculus and fractional differential equations can be obtained by consulting [7], [8], [9], [10] and references cited therein.

This subsection briefly explains the types of fractional derivatives used in the paper. Essentially we use only the Caputo fractional derivatives and Weyl-Liouville fractional derivatives of order $\gamma \in (0, 1]$ defined as follows.

Let $\gamma \in (0,1)$. Then the Caputo fractional derivative $\mathbf{D}_t^\gamma u(t)$ is defined for those functions $u : [0, \infty) \to X$ satisfying that, for every $T > 0$, we have $u_{|(0,T]}(\cdot) \in C((0,T] : X)$, $u(\cdot) - u(0) \in L^1((0,T) : X)$ and $g_{1-\gamma} * (u(\cdot) - u(0)) \in W^{1,1}((0,T) : X)$, by

$$\mathbf{D}_t^\gamma u(t) = \frac{d}{dt}\left[g_{1-\gamma} * \big(u(\cdot) - u(0)\big)\right](t), \quad t \in (0,T];$$

see [7, p. 7] for the notion of Sobolev space $W^{1,1}((0,T) : X)$. The Weyl-Liouville fractional derivative $D_{t,+}^\gamma u(t)$ of order γ is defined for those continuous functions $u : \mathbb{R} \to X$ such that $t \mapsto \int_{-\infty}^t g_{1-\gamma}(t-s)u(s)\,ds$, $t \in \mathbb{R}$ is a well-defined continuously differentiable mapping, by

$$D_{t,+}^\gamma u(t) := \frac{d}{dt}\int_{-\infty}^t g_{1-\gamma}(t-s)u(s)\,ds, \quad t \in \mathbb{R}.$$

Set $\mathbf{D}_t^1 u(t) := (d/dt)u(t)$ and $D_{t,+}^1 u(t) := -(d/dt)u(t)$.

1.2. Multivalued Linear Operators

We need some basic definitions and results about multivalued linear operators in Banach spaces (see [1] for more details in this direction). Suppose that X and Y

are two Banach spaces. By $P(Y)$ we denote the power set of Y. A multivalued map (multimap) $\mathcal{A} : X \to P(Y)$ is said to be a multivalued linear operator, MLO for short, if and only if the following holds:

(i) $D(\mathcal{A}) := \{x \in X : \mathcal{A}x \neq \emptyset\}$ is a linear subspace of X;

(ii) $\mathcal{A}x + \mathcal{A}y \subseteq \mathcal{A}(x+y)$, $x, y \in D(\mathcal{A})$ and $\lambda \mathcal{A}x \subseteq \mathcal{A}(\lambda x)$, $\lambda \in \mathbb{C}$, $x \in D(\mathcal{A})$.

In the case when $X = Y$, we say that $\mathcal{A}$ is an MLO in X. It is well known that for any $x, y \in D(\mathcal{A})$ and $\lambda, \eta \in \mathbb{C}$ with $|\lambda| + |\eta| \neq 0$, we have $\lambda \mathcal{A}x + \eta \mathcal{A}y = \mathcal{A}(\lambda x + \eta y)$. If $\mathcal{A}$ is an MLO, then $\mathcal{A}0$ is a linear manifold in Y and $\mathcal{A}x = f + \mathcal{A}0$ for any $x \in D(\mathcal{A})$ and $f \in \mathcal{A}x$. Define the range $R(\mathcal{A})$ of $\mathcal{A}$ by $R(\mathcal{A}) := \{\mathcal{A}x : x \in D(\mathcal{A})\}$.

Let $\mathcal{A}$ be an MLO in X. Then the resolvent set of $\mathcal{A}$, $\rho(\mathcal{A})$ for short, is defined as the union of those complex numbers $\lambda \in \mathbb{C}$ for which

(i) $X = R(\lambda - \mathcal{A})$;

(ii) $(\lambda - \mathcal{A})^{-1}$ is a single-valued linear continuous operator on X.

The operator $\lambda \mapsto (\lambda - \mathcal{A})^{-1}$ is called the resolvent of $\mathcal{A}$ ($\lambda \in \rho(\mathcal{A})$). Set $R(\lambda : \mathcal{A}) \equiv (\lambda - \mathcal{A})^{-1}$ ($\lambda \in \rho(\mathcal{A})$).

Henceforth, we will employ the following condition:

(P) There exist finite constants $c, M > 0$ and $\beta \in (0, 1]$ such that

$$\Psi := \left\{\lambda \in \mathbb{C} : \Re\lambda \geq -c(|\Im\lambda| + 1)\right\} \subseteq \rho(\mathcal{A})$$

and

$$\|R(\lambda : \mathcal{A})\| \leq M(1 + |\lambda|)^{-\beta}, \quad \lambda \in \Psi.$$

1.3. Lebesgue Spaces with Variable Exponents $L^{p(x)}$

Assume $\emptyset \neq \Omega \subseteq \mathbb{R}$. By $M(\Omega : X)$ we denote the collection of all measurable functions $f : \Omega \to X$; the symbol $M(\Omega)$ stands for the collection of all functions $f \in M(\Omega : \mathbb{C})$ such that $f(x) \in \mathbb{R}$ for all $x \in \Omega$. Furthermore, $\mathcal{P}(\Omega)$

denotes the vector space of all Lebesgue measurable functions $p : \Omega \to [1, \infty]$. For any $p \in \mathcal{P}(\Omega)$ and $f \in M(\Omega : X)$, set

$$\varphi_{p(x)}(t) := \begin{cases} t^{p(x)}, & t \geq 0, \ 1 \leq p(x) < \infty, \\ 0, & 0 \leq t \leq 1, \ p(x) = \infty, \\ \infty, & t > 1, \ p(x) = \infty \end{cases}$$

and

$$\rho(f) := \int_{\Omega} \varphi_{p(x)}(\|f(x)\|) \, dx.$$

We define the Lebesgue space $L^{p(x)}(\Omega : X)$ with variable exponent as follows

$$L^{p(x)}(\Omega : X) := \left\{ f \in M(\Omega : X) : \lim_{\lambda \to 0+} \rho(\lambda f) = 0 \right\}.$$

Then

$$L^{p(x)}(\Omega : X) = \left\{ f \in M(\Omega : X) : \text{ there exists } \lambda > 0 \text{ such that } \rho(\lambda f) < \infty \right\};$$

see [11, p. 73]. For every $u \in L^{p(x)}(\Omega : X)$, we introduce the Luxemburg norm of $u(\cdot)$ in the following manner

$$\|u\|_{p(x)} := \|u\|_{L^{p(x)}(\Omega:X)} := \inf \left\{ \lambda > 0 : \rho(f/\lambda) \leq 1 \right\}.$$

Equipped with the above norm, the space $L^{p(x)}(\Omega : X)$ becomes a Banach one (see e.g., [11, Theorem 3.2.7] for scalar-valued case), coinciding with the usual Lebesgue space $L^p(\Omega : X)$ in the case when $p(x) = p \geq 1$ is a constant function. For any $p \in M(\Omega)$, we set

$$p^- := \operatorname{essinf}_{x \in \Omega} p(x) \quad \text{and} \quad p^+ := \operatorname{esssup}_{x \in \Omega} p(x).$$

Define

$$C_+(\Omega) := \left\{ p \in M(\Omega) : 1 < p^- \leq p(x) \leq p^+ < \infty \text{ for a.e. } x \in \Omega \right\}$$

and

$$D_+(\Omega) := \left\{ p \in M(\Omega) : 1 \leq p^- \leq p(x) \leq p^+ < \infty \text{ for a.e. } x \in \Omega \right\}.$$

Set

$$E^{p(x)}(\Omega : X) := \left\{ f \in L^{p(x)}(\Omega : X) : \text{ for all } \lambda > 0 \text{ we have } \rho(\lambda f) < \infty \right\};$$

$E^{p(x)}(\Omega) \equiv E^{p(x)}(\Omega : \mathbb{C})$. It is well known that $E^{p(x)}(\Omega : X) = L^{p(x)}(\Omega : X)$, provided that $p \in D_+(\Omega)$ (see e.g. [12]).

We will use the following lemma (see e.g. [11, Lemma 3.2.20, (3.2.22); Corollary 3.3.4; Lemma 3.2.8(c)] for the scalar-valued case):

Lemma 1.1. *(i) Let $p, q, r \in \mathcal{P}(\Omega)$, and let*

$$\frac{1}{q(x)} = \frac{1}{p(x)} + \frac{1}{r(x)}, \quad x \in \Omega.$$

Then, for every $u \in L^{p(x)}(\Omega : X)$ and $v \in L^{r(x)}(\Omega)$, we have $uv \in L^{q(x)}(\Omega : X)$ and

$$\|uv\|_{q(x)} \leq 2\|u\|_{p(x)}\|v\|_{r(x)}.$$

(ii) Let Ω be of a finite Lebesgue measure, let $p, q \in \mathcal{P}(\Omega)$, and let $q \leq p$ a.e. on Ω. Then $L^{p(x)}(\Omega : X)$ is continuously embedded in $L^{q(x)}(\Omega : X)$.

(iii) Let $p \in \mathcal{P}(\Omega)$, and let $f_k, f \in M(\Omega : X)$ for all $k \in \mathbb{N}$. If $\lim_{k \to \infty} f_k(x) = f(x)$ for a.e. $x \in \Omega$ and there exists a real-valued function $g \in E^{p(x)}(\Omega)$ such that $\|f_k(x)\| \leq g(x)$ for a.e. $x \in \Omega$, then $\lim_{k \to \infty} \|f_k - f\|_{L^{p(x)}(\Omega:X)} = 0$.

For more details about Lebesgue spaces with variable exponents $L^{p(x)}$, the reader may consult [2], [11]-[12], [14] and [15].

2. GENERALIZED ALMOST PERIODIC AND GENERALIZED ALMOST AUTOMORPHIC FUNCTIONS

Let $1 \leq p < \infty$, and let $f, g \in L^p_{loc}(I : X)$, where $I = \mathbb{R}$ or $I = [0, \infty)$. We define the Stepanov 'metric' by

$$D^p_S[f(\cdot), g(\cdot)] := \sup_{x \in I} \left[\int_x^{x+1} \|f(t) - g(t)\|^p \, dt \right]^{1/p}.$$

The Stepanov norm of $f(\cdot)$ is introduced by $\|f\|_{S^p} := D_S^p[f(\cdot), 0]$. It is said that a function $f \in L_{loc}^p(I : X)$ is Stepanov p-bounded, S^p-bounded shortly, if and only if

$$\|f\|_{S^p} := \sup_{t \in I} \left(\int_t^{t+1} \|f(s)\|^p \, ds \right)^{1/p} = \sup_{t \in I} \left(\int_0^1 \|f(s+t)\|^p \, ds \right)^{1/p} < \infty.$$

Furnished with the above norm, the space $L_S^p(I : X)$ consisted of all S^p-bounded functions is a Banach one. We refer the reader to [1] for the notions of almost periodic functions and Stepanov p-almost periodic functions (see also [13] and [5]).

Let $f : \mathbb{R} \to X$ be continuous. As it is well known, $f(\cdot)$ is called almost automorphic, a.a. for short, if and only if for every real sequence (b_n) there exist a subsequence (a_n) of (b_n) and a map $g : \mathbb{R} \to X$ such that

$$\lim_{n \to \infty} f(t + a_n) = g(t) \quad \text{and} \quad \lim_{n \to \infty} g(t - a_n) = f(t), \tag{1}$$

pointwise for $t \in \mathbb{R}$. If this is the case, $f \in C_b(\mathbb{R} : X)$ and the limit function $g(\cdot)$ must be bounded on $\mathbb{R}$ but not necessarily continuous on $\mathbb{R}$. It is said that $f(\cdot)$ is compactly almost automorphic if and only if the convergence in (1) is uniform on compacts of $\mathbb{R}$. The vector space consisting of all almost automorphic, resp., compactly almost automorphic functions, is denoted by $AA(\mathbb{R} : X)$, resp., $AA_c(\mathbb{R} : X)$. By Bochner's criterion [13], any almost periodic function has to be compactly almost automorphic.

The space of pseudo-almost automorphic functions, denoted by $PAA(\mathbb{R} : X)$, is defined as the direct sum of spaces $AA(\mathbb{R} : X)$ and $PAP_0(\mathbb{R} : X)$, where $PAP_0(\mathbb{R} : X)$ denotes the space consisting of all bounded continuous functions $\Phi : \mathbb{R} \to X$ such that

$$\lim_{r \to \infty} \frac{1}{2r} \int_{-r}^r \|\Phi(s)\| \, ds = 0.$$

Equipped with the sup-norm, the space $PAA(\mathbb{R} : X)$ is a Banach one.

Following G. M. N'Guérékata and A. Pankov [16], we say that a function $f \in L_{loc}^p(\mathbb{R} : X)$ is Stepanov p-almost automorphic, S^p-almost automorphic or S^p-a.a. for short, if and only if for every real sequence (a_n), there exists a subsequence (a_{n_k}) and a function $g \in L_{loc}^p(\mathbb{R} : X)$ such that

$$\lim_{k \to \infty} \int_t^{t+1} \left\| f(a_{n_k} + s) - g(s) \right\|^p ds = 0 \tag{2}$$

and

$$\lim_{k\to\infty} \int_t^{t+1} \left\| g(s - a_{n_k}) - f(s) \right\|^p ds = 0 \qquad (3)$$

for each $t \in \mathbb{R}$; a function $f \in L^p_{loc}([0, \infty) : X)$ is said to be asymptotically Stepanov p-almost automorphic, asymptotically S^p-a.a. shortly, if and only if there exists an S^p-almost automorphic function $g(\cdot)$ and a function $q \in L^p_S([0, \infty) : X)$ such that $f(t) = g(t) + q(t)$, $t \geq 0$ and $\hat{q} \in C_0([0, \infty) : L^p([0, 1] : X))$; any S^p-almost automorphic function $f(\cdot)$ has to be S^p-bounded ($1 \leq p < \infty$); here and hereafter, $\hat{q}(t) := q(t + \cdot)$, $t \geq 0$. The vector space consisting of all S^p-almost automorphic functions, resp., asymptotically S^p-almost automorphic functions, will be denoted by $AAS^p(\mathbb{R} : X)$, resp., $AAAS^p([0, \infty) : X)$.

If $1 \leq p < q < \infty$ and $f(\cdot)$ is (asymptotically) Stepanov q-almost automorphic, then $f(\cdot)$ is (asymptotically) Stepanov p-almost automorphic. Therefore, the (asymptotical) Stepanov p-almost automorphy of $f(\cdot)$ for some $p \in [1, \infty)$ implies the (asymptotical) Stepanov 1-almost automorphy of $f(\cdot)$. As is well known if $f(\cdot)$ is an almost automorphic (a.a.a.) function then $f(\cdot)$ is also S^p-almost automorphic (asymptotically S^p-a.a.) for $1 \leq p < \infty$. However, the converse is false.

A function $f(\cdot)$ is said to be (asymptotically) Stepanov almost periodic (automorphic) if and only if $f(\cdot)$ is (asymptotically) Stepanov 1-almost periodic (automorphic).

3. Generalized Almost Automorphic Type Functions in Lebesgue Spaces with Variable Exponents $L^{p(x)}$

The following notion of Stepanov $p(x)$-boundedness was recently introduced in [1] using a completely different approach from that employed in earlier papers by T. Diagana and M. Zitane (cf. [15, Definition 3.10] and [2, Definition 4.5]):

Definition 1. *Let $p \in \mathcal{P}([0, 1])$, and let $I = \mathbb{R}$ or $I = [0, \infty)$. Then it is said that a function $f \in M(I : X)$ is Stepanov $p(x)$-bounded, $S^{p(x)}$-bounded for short, if and only if $f(\cdot + t) \in L^{p(x)}([0, 1] : X)$ for all $t \in I$, and $\sup_{t \in I} \|f(\cdot + t)\|_{p(x)} <$*

∞, i.e.,

$$\|f\|_{S^{p(x)}} := \sup_{t \in I} \inf \left\{ \lambda > 0 : \int_0^1 \varphi_{p(x)} \left(\frac{\|f(x+t)\|}{\lambda} \right) dx \leq 1 \right\} < \infty.$$

By $L_S^{p(x)}(I : X)$ we denote the vector space consisting of all such functions.

Denote by $\hookrightarrow$ a continuous embedding between normed spaces. Furnished with the norm $\|\cdot\|_{S^{p(x)}}$, the space $L_S^{p(x)}(I : X)$ consisted of all $S^{p(x)}$-bounded functions is a Banach one and we have $L_S^{p(x)}(I : X) \hookrightarrow L_S^1(I : X)$, for any $p \in \mathcal{P}([0,1])$. The space $L_S^{p(x)}(I : X)$ is translation invariant in the sense that, for every $f \in L_S^{p(x)}(I : X)$ and $\tau \in I$, we have $f(\cdot + \tau) \in L_S^{p(x)}(I : X)$.

In [1], we have introduced the concept of (asymptotical) $S^{p(x)}$-almost periodicity as follows:

Definition 2. (i) *Let $p \in \mathcal{P}([0,1])$, and let $I = \mathbb{R}$ or $I = [0, \infty)$. Then it is said that a function $f \in L_S^{p(x)}(I : X)$ is Stepanov $p(x)$-almost periodic, Stepanov $p(x)$-a.p. for short, if and only if the function $\hat{f} : I \to L^{p(x)}([0,1] : X)$ is almost periodic. By $APS^{p(x)}(I : X)$ we denote the vector space consisting of all such functions.*

(ii) *Let $p \in \mathcal{P}([0,1])$, and let $I = [0, \infty)$. Then it is said that a function $f \in L_S^{p(x)}(I : X)$ is asymptotically Stepanov $p(x)$-almost periodic, asymptotically Stepanov $p(x)$-a.p. for short, if and only if the function $\hat{f} : I \to L^{p(x)}([0,1] : X)$ is asymptotically almost periodic. By $AAPS^{p(x)}(I : X)$ we denote the vector space consisting of all such functions; the abbreviation $S_0^{p(x)}([0, \infty) : X)$ will be used to denote the set of all functions $q \in L_S^{p(x)}([0, \infty) : X)$ such that $\hat{q} \in C_0([0, \infty) : L^{p(x)}([0,1] : X))$.*

We know that the space $APS^{p(x)}(I : X)$ is translation invariant in the sense that for every $f \in APS^{p(x)}(I : X)$ and $\tau \in I$, we have $f(\cdot + \tau) \in APS^{p(x)}(I : X)$. A similar statement holds for the space $AAPS^{p(x)}([0, \infty) : X)$.

Now we introduce the concept of $S^{p(x)}$-almost automorphy as follows:

Definition 3. *Let $p \in \mathcal{P}([0,1])$. Then it is said that a function $f \in L_S^{p(x)}(\mathbb{R} : X)$ is Stepanov $p(x)$-almost automorphic, Stepanov $p(x)$-a.a. for short, if and only* if for every real sequence (a_n), there exists a subsequence (a_{n_k}) and a function $g \in L_S^{p(x)}(\mathbb{R} : X)$ such that

$$\lim_{k \to \infty} \left\| f(a_{n_k} + \cdot + t) - g(\cdot + t) \right\|_{L^{p(x)}([0,1]:X)} = 0 \tag{4}$$

and

$$\lim_{k \to \infty} \left\| g(\cdot - a_{n_k} + t) - f(\cdot + t) \right\|_{L^{p(x)}([0,1]:X)} = 0$$

for each $t \in \mathbb{R}$. By $AAS^{p(x)}(\mathbb{R} : X)$ we denote the vector subspace of $L_S^{p(x)}(\mathbb{R} : X)$ consisting of all such functions.

For asymptotical $S^{p(x)}$-almost automorphy, we will use the following notion:

Definition 4. Let $p \in \mathcal{P}([0, 1])$. A function $f \in L_S^{p(x)}([0, \infty) : X)$ is said to be asymptotically Stepanov $p(x)$-almost automorphic, asymptotically $S^{p(x)}$-a.a. shortly, if and only if there exist an $S^{p(x)}$-almost automorphic function $g \in AAS^{p(x)}(\mathbb{R} : X)$ and a function $q \in L_S^{p(x)}([0, \infty) : X)$ such that $f(t) = g(t) + q(t), t \geq 0$ and $\hat{q} \in C_0([0, \infty) : L^{p(x)}([0, 1] : X))$.

It follows immediately from definition that the spaces $AAS^{p(x)}(\mathbb{R} : X)$ and $AAAS^{p(x)}([0, \infty) : X)$ are translation invariant, with the meaning clear. Furthermore, it is simple to check that the notions of Stepanov $p(x)$-boundedness and (asymptotical) Stepanov $p(x)$-almost automorphy are equivalent with those introduced in the previous section, provided that $p(x) \equiv p \geq 1$ is a constant function. Furthermore, the following holds:

Proposition 3.1. Let $p \in \mathcal{P}([0, 1])$ and let $f : \mathbb{R} \to X$ be $S^{p(x)}$-almost periodic. Then $f(\cdot)$ is $S^{p(x)}$-almost automorphic.

Proof. Let (a_n) be a given real sequence. By Bochner's criterion [13], there exists a subsequence (a_{n_k}) of (a_n) and a uniformly continuous bounded function $G : \mathbb{R} \to L^{p(x)}([0, 1] : X)$ such that

$$\lim_{k \to \infty} \sup_{t \in \mathbb{R}} \left\| f(t + a_{n_k} + \cdot) - G(t)(\cdot) \right\|_{L^{p(x)}([0,1]:X)} = 0. \tag{5}$$

It suffices to show that there exists a function $g : \mathbb{R} \to L^{p(x)}([0, 1] : X)$ such that $G(t)(s) = g(t + s)$ for any $t \in \mathbb{R}$ and a.e. $s \in [0, 1]$. We define $g(\cdot) :=$

$G(\lfloor \cdot \rfloor)(\cdot - \lfloor \cdot \rfloor)$. Then it is clear that, for every $t \in \mathbb{Z}$, we have $G(t)(\cdot) = g(t+\cdot)$ a.e. on $[0, 1]$. Suppose that $t \notin \mathbb{Z}$. Since, clearly, $g : \mathbb{R} \to L^{p(x)}([0, 1] : X)$, we only need to prove that $G(t)(s) = G(\lfloor t \rfloor)(t - \lfloor t \rfloor + s)$ for a.e. $s \in (0, \lceil t \rceil - t)$ and $G(t)(s) = G(\lceil t \rceil)(t - \lceil t \rceil + s)$ for a.e. $s \in (\lceil t \rceil - t, 1)$. For the sake of brevity, we will prove the validity of second equality. Since $\int_{\lceil t \rceil - t}^{1} \cdots \leq \int_{0}^{1} \cdots$, (5) implies that

$$\inf\left\{\lambda > 0 : \int_{\lceil t \rceil - t}^{1} \varphi_{p(x)}\left(\frac{\|f(t + a_{n_k} + x) - G(t)(x)\|}{\lambda}\right) dx \leq 1\right\}$$
$$= \inf\left\{\lambda > 0 : \int_{0}^{1+t-\lceil t \rceil} \varphi_{p(x)}\left(\frac{\|f(\lceil t \rceil + a_{n_k} + x) - G(t)(x - \lceil t \rceil + t)\|}{\lambda}\right) dx \leq 1\right\} \to 0,$$

as $k \to \infty$. On the other hand, by (5) and $\int_{0}^{1+t-\lceil t \rceil} \cdots \leq \int_{0}^{1} \cdots$, we have

$$\inf\left\{\lambda > 0 : \int_{0}^{1+t-\lceil t \rceil} \varphi_{p(x)}\left(\frac{\|f(\lceil t \rceil + a_{n_k} + x) - G(t)(x - \lceil t \rceil + t)\|}{\lambda}\right) dx \leq 1\right\} \to 0,$$

as $k \to \infty$. The uniqueness of limits in the space $L^{p(x)}([0, 1+t-\lceil t \rceil] : X)$ yields the required equality. □

Making use of Proposition 3.1 and [1, Proposition 3.12], we immediately get:

Proposition 3.2. *Let $p \in \mathcal{P}([0, 1])$ and let $f : [0, \infty) \to X$ be asymptotically $S^{p(x)}$-almost periodic. Then $f(\cdot)$ is asymptotically $S^{p(x)}$-almost automorphic.*

Remark 1. *Albeit the introduced function spaces are translation invariant, it is our duty to say that the space $L_S^{p(x)}(I : X)$ consisting of all Stepanov $p(x)$-bounded functions, which is crucial to our work, has numerous peculiarities that we will not discuss here. It is also clear that our basic assumption $p \in \mathcal{P}([0, 1])$ can be replaced with the assumption that $p \in \mathcal{P}([0, a])$ for all (some) $a > 0$, so that Definition 1 can be modified and used for introducing a great deal of different spaces of (asymptotically) Stepanov almost automorphic functions with variable exponents (cf. also Definition 2-Definition 4); we will not follow this approach here for the sake of brevity.*

Assume that $p \in \mathcal{P}([0, 1])$, resp. $p \in D_+([0, 1])$, and $1 \leq p^- \leq p(x) \leq p^+ < \infty$ for a.e. $x \in [0, 1]$. Then Lemma 1.1(ii) implies that $L^\infty(\mathbb{R} : X) \hookrightarrow L_S^{p(x)}(\mathbb{R} : X) \hookrightarrow L_S^1(\mathbb{R} : X)$, resp., $L_S^{p^+}(\mathbb{R} : X) \hookrightarrow L_S^{p(x)}(\mathbb{R} : X) \hookrightarrow L_S^{p^-}(\mathbb{R} :$

X). Therefore, a similar line of reasoning to the almost periodic case shows that the following theorem holds true; for the sake of completeness, we will prove only the second part of (i), $AAAS^{p(x)}([0,\infty):X) \hookrightarrow AAAS^1([0,\infty):X)$:

Theorem 3.3. (i) Let $p \in \mathcal{P}([0,1])$. Then $AAS^{p(x)}(\mathbb{R}:X) \hookrightarrow AAS^1(\mathbb{R}:X)$ and
$AAAS^{p(x)}([0,\infty):X) \hookrightarrow AAAS^1([0,\infty):X)$.

(ii) Let $p \in D_+([0,1])$ and $1 \leq p^- \leq p(x) \leq p^+ < \infty$ for a.e. $x \in [0,1]$. Then $AAS^{p^+}(\mathbb{R}:X) \hookrightarrow AAS^{p(x)}(\mathbb{R}:X) \hookrightarrow AAS^{p^-}(\mathbb{R}:X)$ and $AAAS^{p^+}([0,\infty):X) \hookrightarrow AAAS^{p(x)}([0,\infty):X) \hookrightarrow AAAS^{p^-}([0,\infty):X)$.

(iii) If $p, q \in \mathcal{P}([0,1])$ and $p \leq q$ a.e. on $[0,1]$, then $AAS^{q(x)}(\mathbb{R}:X) \hookrightarrow AAS^{p(x)}(\mathbb{R}:X)$ and $AAAS^{q(x)}([0,\infty):X) \hookrightarrow AAAS^{p(x)}([0,\infty):X)$.

Proof. Let $f \in AAAS^{p(x)}([0,\infty):X)$. By definition, there exist an $S^{p(x)}$-almost automorphic function $g(\cdot)$ and a function $q \in L_S^{p(x)}([0,\infty):X)$ such that $f(t) = g(t) + q(t), t \geq 0$ and $\hat{q} \in C_0([0,\infty):L^{p(x)}([0,1]:X))$. It is clear that $g(\cdot)$ is S^1-almost automorphic and $\hat{q} \in C_0([0,\infty):L^1([0,1]:X))$ because $L^{p(x)}([0,1]:X) \hookrightarrow L^1([0,1]:X)$; therefore, $f \in AAAS^1([0,\infty):X)$. Using the fact that $L_S^{p(x)}([0,\infty):X) \hookrightarrow L_S^1([0,\infty):X)$, it readily follows that there is a finite constant $c > 0$, independent of $f(\cdot)$, such that

$$\|f\|_{L_S^{p(x)}([0,\infty):X)} \leq c\|f\|_{L_S^1([0,\infty):X)}.$$

This completes the proof. □

Problem. In [1], we have proved the following: If $p \in D_+([0,1])$, then

$$L^\infty(\mathbb{R}:X) \cap APS^{p(x)}(\mathbb{R}:X) = L^\infty(\mathbb{R}:X) \cap APS^1(\mathbb{R}:X)$$

and

$$L^\infty([0,\infty):X) \cap AAPS^{p(x)}([0,\infty):X) = L^\infty([0,\infty):X) \cap AAPS^1([0,\infty):X).$$

The proof given in the above-mentioned paper does not work for almost automorphy with variable exponent. Thus, we would like to ask whether the assumption $p \in D_+([0,1])$ implies

$$L^\infty(\mathbb{R}:X) \cap AAS^{p(x)}(\mathbb{R}:X) = L^\infty(\mathbb{R}:X) \cap AAS^1(\mathbb{R}:X)$$

and
$$L^\infty([0,\infty):X) \cap AAAS^{p(x)}([0,\infty):X) = L^\infty([0,\infty):X) \cap AAAS^1([0,\infty):X)?$$

The subsequent lemma can be deduced following the lines of proof of [1, Proposition 3.5]:

Lemma 3.1. *Assume that $p \in \mathcal{P}([0,1])$ and $q \in C_0([0,\infty):X)$. Then $q \in L_S^{p(x)}([0,\infty):X)$ and $\hat{q} \in C_0([0,\infty):L^{p(x)}([0,1]:X))$.*

For the sequel, it is worth noting that due to an elementary line of reasoning, we have $AA_c(\mathbb{R}:X) = AAS^\infty(\mathbb{R}:X) \cap C_b(\mathbb{R}:X)$. Hence, the function $f(\cdot)$ cannot belong to the class $AAS^{p(x)}(\mathbb{R}:X)$ if $f(\cdot)$ is almost automorphic, not compactly almost automorphic, and $p(x) \equiv \infty$, $x \in [0,1]$. Similarly, if $g(\cdot)$ is almost automorphic, not compactly almost automorphic and $q \in C_0([0,\infty):X)$, then the function $f \equiv g + q$ cannot belong to the class $AAAS^\infty([0,\infty):X)$. In the following proposition, we will find a simple sufficient condition on $p \in \mathcal{P}([0,1])$ ensuring that an (asymptotically) almost automorphic function is (asymptotically) $S^{p(x)}$-almost automorphic:

Proposition 3.4. *Let $p \in \mathcal{P}([0,1])$, let $f : \mathbb{R} \to X$ be almost automorphic, resp., $f : [0,\infty) \to X$ be asymptotically almost automorphic, and let*

$$\int_0^1 \lambda^{p(x)} \, dx < \infty \text{ for all } \lambda > 0. \tag{6}$$

Then $f(\cdot)$ is $S^{p(x)}$-almost automorphic, resp., $f(\cdot)$ is asymptotically $S^{p(x)}$-almost automorphic.

Proof. The argumentation used in the almost periodic case shows that $f(\cdot)$ is $S^{p(x)}$-bounded and $\|f\|_{L_S^{p(x)}} \leq \|f\|_\infty$. Let (b_n) be a given real sequence. Then there exist a subsequence (a_n) of (b_n) and a map $g : \mathbb{R} \to X$ such that (1) holds, pointwise for $t \in \mathbb{R}$. It is well known that $g \in L^\infty(\mathbb{R}:X)$ and, by [1, Proposition 3.6(i)], $g \in L_S^{p(x)}(\mathbb{R}:X)$. Due to (6), we get that the Lebesgue measure of the set $\{x \in [0,1] : p(x) = \infty\}$ is equal to zero and therefore any essentially bounded function $h : \mathbb{R} \to X$ satisfies that for every $\lambda > 0$, we have $\int_0^1 \varphi_{p(x)}(\|h(x)\|/\lambda) \, dx < \infty$. Using this fact, we can apply Lemma 1.1(iii) in order to see that

$$\lim_{n \to \infty} \|f(t + a_n + \cdot) - g(t + \cdot)\|_{L^{p(x)}([0,1]:X)} = 0$$

and
$$\lim_{n\to\infty} \|g(t + \cdot - a_n) - f(t + \cdot)\|_{L^{p(x)}([0,1]:X)} = 0,$$

pointwise for $t \in \mathbb{R}$. This completes the proof of proposition for $S^{p(x)}$-almost automorphy; the corresponding result for asymptotical $S^{p(x)}$-almost automorphy follows by combining this and Lemma 3.1. □

Now we will continue our analyses from [1, Example 3.11]:

Example 1. *Set* $\text{sign}(0) := 0$. *Then, for every almost periodic function* $f : \mathbb{R} \to \mathbb{R}$, *we know that the function* $F(\cdot) := \text{sign}(f(\cdot))$ *is Stepanov* $p(x)$-*almost periodic for any* $p \in D_+([0,1])$ *as well as that the function* $F(\cdot)$ *is Stepanov* $p(x)$-*bounded for any* $p \in \mathcal{P}([0,1])$; *see [1]. By Proposition 3.1, we have that* $F \in AAS^{p(x)}(\mathbb{R} : \mathbb{C})$ *for any* $p \in D_+([0,1])$.

In [1], we have further analyze the special case that $f(x) := \sin x + \sin\sqrt{2}x$, $x \in \mathbb{R}$ *and* $p(x) := 1 - \ln x$, $x \in [0,1]$, *showing that* $F \notin APS^{p(x)}(\mathbb{R} : \mathbb{C})$. *Now we will verify that* $F \notin AAS^{p(x)}(\mathbb{R} : \mathbb{C})$. *For this it is sufficient to construct a real sequence* (a_n) *such that, for every* $n \in \mathbb{N}$ *and every* $\lambda \in (0, 2/e)$, *we have*

$$\int_0^1 \left|\frac{F(x + a_{2n}) - F(x + a_{2n-1})}{\lambda}\right|^{1-\ln x} dx = \infty;$$

(see (4) with $t = 0$, *and observe that in this case the sequence* $F(a_{n_k} + \cdot)_{k \in \mathbb{N}}$ *needs to be a Cauchy one in* $L^{p(x)}[0,1]$). *But we have proved that for any* $\lambda \in (0, 2/e)$, *any* $l > 0$, *any interval* $I \subseteq \mathbb{R} \setminus \{0\}$ *of length* $l > 0$ *and any* $\tau \in I$, *there exists* $t \in \mathbb{R}$ *such that*

$$\int_0^1 \left(\frac{1}{\lambda}\right)^{1-\ln x} |F(x + t + \tau) - F(x + t)|^{1-\ln x} dx = \infty. \tag{7}$$

Let $\lambda \in (0, 2/e)$ *be arbitrarily chosen,* $l = 1$, $I = I_n = [n, n+1]$ *and* $\tau = n$. *Then we can find* $t = t_n$ *such that (7) holds, so that the claim follows by plugging* $a_{2n-1} := t_n$ *and* $a_{2n} := t_n + n$ ($n \in \mathbb{N}$).

In the case that $f(x) := \sin x$, $x \in \mathbb{R}$, *we have also proved that the function* $F(\cdot)$ *is* $S^{p(x)}$-*almost periodic for any* $p \in \mathcal{P}([0,1])$. *By Proposition 3.1, we have that* $F(\cdot)$ *is* $S^{p(x)}$-*almost automorphic for any* $p \in \mathcal{P}([0,1])$.

To the best of the authors' knowledge, in the existing literature concerning Stepanov almost automorphic functions, the authors have examined only such functions that are Stepanov p-almost automorphic for any exponent $p \in [1, \infty)$, and therefore, Stepanov $p(x)$-almost automorphic for any function $p \in D_+([0, 1])$ (cf. Theorem 3.3(ii)). Therefore, it is natural to ask whether there exists a Stepanov almost automorphic function that is not Stepanov p-almost automorphic for a certain exponent $p \in (1, \infty)$. The answer is affirmative and, without going into the full problematic concerning this and similar questions, we would like to recall that H. Bohr and E. Følner have constructed, for any given number $p > 1$, a Stepanov almost periodic function defined on the whole real axis that is Stepanov p-bounded and not Stepanov p-almost periodic (see [17, Example, pp. 70-73]). This function, denoted here by $f(\cdot)$, is clearly Stepanov almost automorphic and now we will prove that $f(\cdot)$ cannot be Stepanov p-almost automorphic. Consider, for simplicity, the case that $h_1 = 2$ in the afore-mentioned theorem and suppose the contrary. Then it is well known that the mapping $\hat{f} : \mathbb{R} \to L^p([0, 1] : X)$ is compactly almost automorphic. Since the class of almost automorphic functions coincides with the class of Levitan N-almost periodic functions (see e.g., [13, p. 111] and [18, pp. 53-54]), for every $\epsilon > 0$ and $N > 0$, there exists a finite number $L > 0$ such that any interval $I \subseteq \mathbb{R}$ contains a number $\tau \in I$ such that $\|\hat{f}(t \pm \tau) - \hat{f}(t)\|_{L^p([0,1]:X)} < \epsilon$. Especially, with $\epsilon > 0$ arbitrarily small and $N = 3/2$, we get the existence of a finite number $L > 0$ such that any interval $I \subseteq \mathbb{R} \setminus [-1, 1]$ contains a number $\tau \in I$ such that

$$\int_x^{x+1} |f(s+\tau) - f(s)|^p \, ds < \epsilon^p, \quad x \in [-3/2, 3/2].$$

With $x = -3/2$, this implies

$$\int_{-3/2}^{3/2} |f(s+\tau) - f(s)|^p \, ds < 2\epsilon^p,$$

which is in contradiction with the estimate $\int_{-3/2}^{3/2} |f(s+\tau) - f(s)|^p \, ds \geq 2^{-p}$ (see [17, p. 73, l.-9 - l.-4]).

4. GENERALIZED TWO-PARAMETER ALMOST AUTOMORPHIC TYPE FUNCTIONS AND COMPOSITION PRINCIPLES

Suppose that $(Y, \|\cdot\|_Y)$ is a complex Banach space, as well as that $I = \mathbb{R}$ or $I = [0, \infty)$. By $C_0([0, \infty) \times Y : X)$ we denote the space consisting of all continuous functions $h : [0, \infty) \times Y \to X$ such that $\lim_{t \to \infty} h(t, y) = 0$ uniformly for y in any compact subset of Y. A continuous function $f : I \times Y \to X$ is called uniformly continuous on bounded sets, uniformly for $t \in I$ if and only if for every $\epsilon > 0$ and every bounded subset K of Y there exists a number $\delta_{\epsilon, K} > 0$ such that $\|f(t, x) - f(t, y)\| \leq \epsilon$ for all $t \in I$ and all $x, y \in K$ satisfying that $\|x - y\|_Y \leq \delta_{\epsilon, K}$. If $f : I \times Y \to X$, set $\hat{f}(t, y) := f(t + \cdot, y), t \geq 0, y \in Y$.

We need to recall the following well-known definition (see e.g., [13] and [5] for more details):

Definition 5. *Let $1 \leq p < \infty$.*

(i) *A jointly continuous function $f : \mathbb{R} \times Y \to X$ is said to be almost automorphic if and only if for every sequence of real numbers (s'_n) there exists a subsequence (s_n) such that*

$$G(t, y) := \lim_{n \to \infty} F(t + s_n, y)$$

is well defined for each $t \in \mathbb{R}$ and $y \in Y$, and

$$\lim_{n \to \infty} G(t - s_n, y) = F(t, y)$$

for each $t \in \mathbb{R}$ and $y \in Y$. The vector space consisting of such functions will be denoted by $AA(\mathbb{R} \times Y : X)$.

(ii) *A bounded continuous function $f : \mathbb{R} \times Y \to X$ is said to be pseudo-almost automorphic if and only if $F = G + \Phi$, where $G \in AA(\mathbb{R} \times Y : X)$ and $\Phi \in PAP_0(\mathbb{R} \times Y : X)$; here, $PAP_0(\mathbb{R} \times Y : X)$ denotes the space consisting of all continuous functions $\Phi : \mathbb{R} \times Y \to X$ such that $\{\Phi(t, y) : t \in \mathbb{R}\}$ is bounded for all $y \in Y$, and*

$$\lim_{r \to \infty} \frac{1}{2r} \int_{-r}^{r} \|\Phi(s, y)\| \, ds = 0,$$

uniformly in $y \in Y$. The vector space consisting of such functions will be denoted by $PAA(\mathbb{R} \times Y : X)$.

We introduce the notions of a Stepanov two-parameter $p(x)$-almost automorphic function and an asymptotically Stepanov two-parameter $p(x)$-almost automorphic function as follows:

Definition 6. *Let $p \in \mathcal{P}([0,1])$, and let $f : \mathbb{R} \times Y \to X$ be such that for each $y \in Y$ we have $f(\cdot, y) \in L_S^{p(x)}(\mathbb{R} : X)$. Then we say that $f(\cdot, \cdot)$ is Stepanov $p(x)$-almost automorphic if and only if for every $y \in Y$ the mapping $f(\cdot, y)$ is $S^{p(x)}$-almost automorphic; that is, for any real sequence (a_n) there exist a subsequence (a_{n_k}) of (a_n) and a map $g : \mathbb{R} \times Y \to X$ such that $g(\cdot, y) \in L_S^{p(x)}(\mathbb{R} : X)$ for all $y \in Y$ as well as that:*

$$\lim_{k \to \infty} \left\| f(t + a_{n_k} + \cdot, y) - g(t + \cdot, y) \right\|_{L^{p(x)}[0,1]} = 0$$

and

$$\lim_{k \to \infty} \left\| g(t + \cdot - a_{n_k}, y) - f(t + \cdot, y) \right\|_{L^{p(x)}[0,1]} = 0$$

for each $t \in \mathbb{R}$ and for each $y \in Y$. We denote by $AAS^{p(x)}(\mathbb{R} \times Y : X)$ the vector space consisting of all such functions.

Definition 7. *Let $p \in \mathcal{P}([0,1])$. A function $f : [0, \infty) \times Y \to X$ is said to be asymptotically $S^{p(x)}$-almost automorphic if and only if $\hat{f} : [0, \infty) \times Y \to L^{p(x)}([0,1] : X)$ is asymptotically almost automorphic. The collection of such functions will be denoted by $AAAS^{p(x)}([0, \infty) \times Y : X)$.*

The following well-known result of Fan et al. [4] is reformulated here for Stepanov $p(x)$-almost automorphy:

Theorem 4.1. *Assume that $p \in \mathcal{P}([0,1])$, and $f \in AAS^{p(x)}(\mathbb{R} \times Y : X)$. If there exists a constant $L > 0$ such that for all $x, y \in L_S^{p(x)}(\mathbb{R} : Y)$*

$$\left\| f(t + \cdot, x(\cdot)) - f(t + \cdot, y(\cdot)) \right\|_{L^{p(x)}([0,1]:X)} \leq L_Y \left\| x(\cdot) - y(\cdot) \right\|_{L^{p(x)}([0,1]:Y)}$$

then for each $x \in AAS^p(\mathbb{R} : Y)$ with relatively compact range in Y one has that $f(\cdot, x(\cdot)) \in AAS^p(\mathbb{R} : X)$.

The following result generalizes the one established by Ding et al. [3] (see e.g., [13, pp. 134-138]). The proof is similar and therefore omitted:

Theorem 4.2. *Let $I = \mathbb{R}$, and let $p \in \mathcal{P}([0,1])$. Suppose that the following conditions hold:*

(i) $f \in AAS^{p(x)}(I \times Y : X)$ *and there exist a function* $r \in \mathcal{P}([0,1])$ *such that* $r(\cdot) \geq \max(p(\cdot), p(\cdot)/p(\cdot) - 1)$ *and a function* $L_f \in L_S^{r(x)}(I)$ *such that:*

$$\|f(t,x) - f(t,y)\| \leq L_f(t)\|x-y\|_Y, \quad t \in I, \ x, \ y \in Y; \quad (8)$$

(ii) $u \in AAS^{p(x)}(I : Y)$, *and there exists a set* $\mathrm{E} \subseteq I$ *with* $m(\mathrm{E}) = 0$ *such that* $K := \{u(t) : t \in I \setminus \mathrm{E}\}$ *is relatively compact in* Y; *here,* $m(\cdot)$ *denotes the Lebesgue measure.*

Define $q \in \mathcal{P}([0,1])$ by $q(x) := p(x)r(x)/p(x) + r(x)$, if $x \in [0,1]$ and $r(x) < \infty$; $q(x) := p(x)$, if $x \in [0,1]$ and $r(x) = \infty$. Then $q(x) \in [1, p(x))$ for $x \in [0,1], r(x) < \infty$ and $f(\cdot, u(\cdot)) \in AAS^{q(x)}(I : X)$.

Concerning asymptotical two-parameter Stepanov $p(x)$-almost automorphy, we can deduce the following composition principle with $X = Y$; see the proofs of [5, Proposition 2.7.3, Proposition 2.7.4] for more details:

Proposition 4.3. *Let $I = [0, \infty)$, and let $p \in \mathcal{P}([0,1])$. Suppose that the following conditions hold:*

(i) $g \in AAS^{p(x)}(I \times X : X)$, *there exist a function* $r \in \mathcal{P}([0,1])$ *such that* $r(\cdot) \geq \max(p(\cdot), p(\cdot)/p(\cdot) - 1)$ *and a function* $L_g \in L_S^{r(x)}(I)$ *such that* (8) *holds with the function* $f(\cdot, \cdot)$ *replaced by the function* $g(\cdot, \cdot)$ *therein.*

(ii) $v \in AAS^{p(x)}(I : X)$, *and there exists a set* $\mathrm{E} \subseteq I$ *with* $m(\mathrm{E}) = 0$ *such that* $K = \{v(t) : t \in I \setminus \mathrm{E}\}$ *is relatively compact in X.*

(iii) $f(t,x) = g(t,x) + q(t,x)$ *for all* $t \geq 0$ *and* $x \in X$, *where* $\hat{q} \in C_0([0,\infty) \times X : L^{q(x)}([0,1] : X))$ *with* $q(\cdot)$ *defined as above.*

(iv) $u(t) = v(t) + \omega(t)$ *for all* $t \geq 0$, *where* $\hat{\omega} \in C_0([0,\infty) : L^{p(x)}([0,1] : X))$.

(v) *There exists a set* $E' \subseteq I$ *with* $m(E') = 0$ *such that* $K' = \{u(t) : t \in I \setminus E'\}$ *is relatively compact in* X.

Then $f(\cdot, u(\cdot)) \in AAAS^{q(x)}(I : X)$.

5. GENERALIZED (ASYMPTOTICAL) ALMOST AUTOMORPHY IN LEBESGUE SPACES WITH VARIABLE EXPONENTS $L^{p(x)}$: ACTIONS OF CONVOLUTION PRODUCTS AND SOME APPLICATIONS

We start this section by stating the following generalization of [6, Proposition 5] (the reflexion at zero keeps the spaces of Stepanov p-almost automorphic functions unchanged, which may or may not be the case with the spaces of Stepanov $p(x)$-almost automorphic functions):

Proposition 5.1. *Suppose that* $p \in D_+([0,1])$, $q \in \mathcal{P}([0,1])$, $1/p(x) + 1/q(x) = 1$ *and* $(R(t))_{t>0} \subseteq L(X,Y)$ *is a strongly continuous operator family satisfying that* $M := \sum_{k=0}^{\infty} \|R(\cdot + k)\|_{L^{q(x)}[0,1]} < \infty$. *If* $\breve{g} : \mathbb{R} \to X$ *is* $S^{p(x)}$-*almost automorphic, then the function* $G : \mathbb{R} \to Y$, *given by*

$$G(t) := \int_{-\infty}^{t} R(t-s)g(s)\,ds, \quad t \in \mathbb{R}, \tag{9}$$

is well-defined and almost automorphic.

Proof. The proof of theorem is very similar to that of the above-mentioned proposition since the Hölder inequality holds in our framework (see Lemma 1.1(ii)) and any element of $L^{p(x)}([0,1] : X)$ is absolutely continuous with respect to the norm $\|\cdot\|_{L^{p(x)}}$ (see [12, Definition 1.12, Theorem 1.13]), which clearly implies that the translation mapping $t \mapsto \breve{g}(\cdot - t) \in L^{p(x)}([0,1] : X)$, $t \in \mathbb{R}$ is continuous (we need this fact for proving the continuity of mapping $F_k(\cdot)$ appearing in the proof of [6, Proposition 5], $k \in \mathbb{N}$). The remaining part of the proof can be given by copying the corresponding part of the proof of the above-mentioned proposition. □

In general case $p \in \mathcal{P}([0,1])$, there are elements of $L^{p(x)}([0,1])$ that are not absolutely continuous with respect to the norm $\|\cdot\|_{L^{p(x)}}$; see e.g., [19, p. 602]. In this case, Proposition 5.1 continues to hold if we impose the condition on continuity of mapping $t \mapsto \breve{g}(\cdot - t) \in L^{p(x)}([0,1] : X)$, $t \in \mathbb{R}$ in place of condition $p \in D_+([0,1])$.

Proposition 5.1 can be simply incorporated in the study of existence and uniqueness of almost periodic solutions of the following abstract Cauchy differential inclusion of first order

$$u'(t) \in \mathcal{A}u(t) + g(t), \quad t \in \mathbb{R} \tag{10}$$

and the following abstract Cauchy relaxation differential inclusion

$$D_{t,+}^{\gamma} u(t) \in -\mathcal{A}u(t) + g(t), \, t \in \mathbb{R}, \tag{11}$$

where $\mathcal{A}$ is an MLO satisfying the condition (P), $D_{t,+}^{\gamma}$ denotes the Weyl-Liouville fractional derivative of order $\gamma \in (0,1)$ and $g : \mathbb{R} \to X$ satisfies certain assumptions; see [1] and [5] for further information.

In the following proposition, we state some invariance properties of generalized asymptotical almost automorphy in Lebesgue spaces with variable exponents $L^{p(x)}$ under the action of finite convolution products. This proposition generalizes [6, Proposition 6] provided that $p > 1$ in its formulation.

Proposition 5.2. *Suppose that* $p, q \in C_+([0,1])$, $1/p(x) + 1/q(x) = 1$ *and* $(R(t))_{t>0} \subseteq L(X)$ *is a strongly continuous operator family satisfying that, for every* $t \geq 0$, *we have that* $m_t := \sum_{k=0}^{\infty} \|R(\cdot + t + k)\|_{L^{q(x)}[0,1]} < \infty$. *Suppose, further, that* $\breve{g} : \mathbb{R} \to X$ *is* $S^{p(x)}$-*almost automorphic,* $q \in L_S^{p(x)}([0,\infty) : X)$ *and* $f(t) = g(t) + q(t), t \geq 0$. *Let* $r_1, r_2 \in \mathcal{P}([0,1])$ *and the following hold:*

(i) *For every* $t \geq 0$, *the mapping* $x \mapsto \int_0^{t+x} R(t+x-s)q(s)\,ds$, $x \in [0,1]$ *belongs to the space* $L^{r_1(x)}([0,1] : X)$ *and we have*

$$\lim_{t \to +\infty} \left\| \int_0^{t+x} R(t+x-s)q(s)\,ds \right\|_{L^{r_1(x)}[0,1]} = 0. \tag{12}$$

(ii) *For every* $t \geq 0$, *the mapping* $x \mapsto m_{t+x}$, $x \in [0,1]$ *belongs to the space* $L^{r_2(x)}[0,1]$ *and we have*

$$\lim_{t \to +\infty} |m_{t+x}|_{L^{r_2(x)}[0,1]} = 0.$$

Then the function $H(\cdot)$, *given by* $H(t) := \int_0^t R(t-s)f(s)\,ds$, $t \geq 0$, *is well defined, bounded and belongs to the class* $AAS^{p(x)}(\mathbb{R} : X) + S_0^{r_1(x)}([0,\infty) : X) + S_0^{r_2(x)}([0,\infty) : X)$, *with the meaning clear.*

Proof. Define $G(\cdot)$ by (9) and $F(\cdot)$ by [

$$F(t) := \int_0^t R(t-s)q(s)\,ds - \int_t^\infty R(s)g(t-s)\,ds := F_1(t) + F_2(t), \quad t \geq 0.$$

It can be simply shown that the function $F_1(\cdot)$ is well defined and bounded because $q(\cdot)$ is $S^{p(x)}$-bounded and $m_0 < \infty$; cf. the proof of [1, Proposition 3.14]. Furthermore, the integral $\int_t^\infty R(s)g(t-s)\,ds = F_2(t)$ is well defined for all $t \geq 0$, which follows by applying Lemma 1.1(ii):

$$\left\| \int_t^\infty R(s)g(t-s)\,ds \right\| = \left\| \int_t^\infty R(s)\check{g}(s-t)\,ds \right\|$$

$$\leq \sum_{k=0}^\infty \int_{t+k}^{t+k+1} \|R(s)\| \|\check{g}(s-t)\|\,ds = \sum_{k=0}^\infty \int_0^1 \|R(s+t+k)\|\|\check{g}(s+k)\|\,ds$$

$$\leq 2\|\check{g}\|_{L_S^{p(x)}(\mathbb{R}:X)} \sum_{k=0}^\infty \|R(t+k+\cdot)\|_{L^{q(x)}[0,1]} \leq 2\|\check{g}\|_{L_S^{p(x)}(\mathbb{R}:X)} m_t < \infty, \quad t \geq 0.$$

Since $H(t) = G(t) + F(t)$ for all $t \geq 0$, we get that the function $H(\cdot)$ is well defined and bounded; due to Proposition 5.1, it remains to be shown that the mapping $\hat{F}_i : [0,\infty) \to L^{r_i(x)}([0,1] : X)$ is in class $C_0([0,\infty) : L^{r_i(x)}([0,1] : X))$ for $i = 1, 2$. Let $k \in \mathbb{N}_0$. For the continuity of mapping $t \mapsto F_{k,2}(t) := \int_{t+k}^{t+k+1} R(s)g(t-s)\,ds$, $t \geq 0$, let us assume that (t_n) is a sequence of positive reals converging to some fixed number $t \geq 0$. Having in mind Lemma 1.1(ii), we obtain that

$$\|F_{k,2}(t_n) - F_{k,2}(t)\| \leq 2\|\check{g}\|_{L_S^{p(x)}(\mathbb{R}:X)} \|R(t_n+k+\cdot) - R(t+k+\cdot)\|_{L^{q(x)}[0,1]}, \quad n \in \mathbb{N},$$

so that the claimed assertion follows by applying [12, Theorem 1.13] (observe that we need the condition $p, q \in C_+([0,1])$ here). Using the condition $m_t < \infty$ as well as the Weierstrass criterion (see also the proof of [6, Proposition 5]), we get that the mapping $F_2(\cdot)$ is continuous. The continuity of mapping $t \mapsto \int_0^t R(t-s)q(s)\,ds$, $t \geq 0$ can be shown similarly. By [1, Proposition 3.7(iii)], we get that the mapping $\hat{F}_i : [0,\infty) \to L^{r_i(x)}([0,1] : X)$ is continuous for $i = 1, 2$. Taking into account (i)-(ii) and the computation used above for proving the boundedness of function $F_2(\cdot)$, we easily get that $\lim_{t \to +\infty} \|F_i(t+\cdot)\|_{L^{r_i(x)}([0,1]:X)} = 0$ for $i = 1, 2$. The proof of the proposition is thereby complete. $\square$

The next example exhibits the use of ergodic Stepanov components with variable exponents (see also Example 3 below):

Example 2. *Suppose that $p(x) \equiv r_2(x) \equiv p > 1$, $q(x) \equiv p/p - 1 > 1$ ($x \in [0, 1]$), $r_1 \in \mathcal{P}([0, 1])$, $(R(t))_{t \geq 0}$ is strongly continuous, exponentially decaying, $g : \mathbb{R} \to X$ is S^p-almost periodic, $q : [0, \infty) \to X$ is S^p-bounded, (12) holds but*

$$\lim_{t \to +\infty} \left\| \int_0^{t+x} R(t + x - s) q(s) \, ds \right\|_{L^p[0,1]} \neq 0.$$

Then the function $H(\cdot)$ is bounded and belongs to the class $AAAS^{p(\cdot)}([0, \infty) : X) + S_0^{r_1(x)}([0, \infty) : X)$ but not to the class $AAAS^{p(x)}([0, \infty) : X)$; see also [20, Remark 2.14(i)].

We can simply apply Proposition 5.2 in the analysis of existence and uniqueness of asymptotically $S^{r(x)}$-almost automorphic solutions for a wide class of abstract Volterra integro-differential equations and inclusions. For example, Proposition 5.2 is applicable to the analysis of asymptotically $S^{r(x)}$-almost automorphic solutions of the following abstract integro-differential inclusion:

$$\left[u(\cdot) - (g_{\zeta+1+i} * f)(\cdot) C x \right] + \sum_{j=1}^{n-1} c_j g_{\alpha_n - \alpha_j} * \left[u(\cdot) - (g_{\zeta+1+i} * f)(\cdot) C x \right]$$
$$+ \sum_{j \in \mathbb{N}_{n-1} \setminus D_i} c_j \left[g_{\alpha_n - \alpha_j + i + \zeta + 1} * f \right](\cdot) C x \in \mathcal{A}\left[g_{\alpha_n - \alpha} * u \right](\cdot),$$

where $\zeta \geq 0$ is appropriately chosen, $C \in L(X)$ commutes with $\mathcal{A}$, $c_j \geq 0$ for $1 \leq j \leq n - 1$, $0 \leq \alpha_1 < \cdots < \alpha_n$, $0 \leq \alpha < \alpha_n$ and $f(\cdot)$ satisfies the requirements of Proposition 5.2 (cf. [21] for the notion of sets D_i and more details on the subject).

In what follows, we will briefly explain how one can apply Proposition 5.2 in the study of qualitative analysis of solutions of the following fractional relaxation inclusion

$$(\text{DFP})_{f,\gamma} : \begin{cases} \mathbf{D}_t^\gamma u(t) \in \mathcal{A} u(t) + f(t), \ t > 0, \\ u(0) = x_0, \end{cases}$$

where $\mathbf{D}_t^\gamma$ denotes the Caputo fractional derivative of order $\gamma \in (0, 1]$, $x_0 \in X$ and $f : [0, \infty) \to X$ satisfies certain properties. Let $(S_\gamma(t))_{t>0}$ and $(P_\gamma(t))_{t>0}$

be the operator families defined in [1]. Then we have the existence of two finite constants $M_1 > 0$ and $M_2 > 0$ such that

$$\|S_\gamma(t)\| + \|P_\gamma(t)\| \le M_1 t^{\gamma(\beta-1)}, \quad t > 0 \tag{13}$$

and

$$\|S_\gamma(t)\| \le M_2 t^{-\gamma}, \ t \ge 1, \quad \|P_\gamma(t)\| \le M_2 t^{-2\gamma}, \ t \ge 1. \tag{14}$$

Set $R_\gamma(t) := t^{\gamma-1} P_\gamma(t)$, $t > 0$. By a mild solution of (DFP)$_{f,\gamma}$, we mean any function $u \in C([0, \infty) : X)$ satisfying that

$$u(t) = S_\gamma(t) x_0 + \int_0^t R_\gamma(t-s) f(s)\, ds, \quad t \ge 0.$$

The estimates (13)-(14) and the representation formula for $u(\cdot)$ are crucial for applications of Proposition 5.2. We provide below an illustrative example:

Example 3. Let $x_0 \in X$ belong to the domain of continuity of $(T(t))_{t>0}$, i.e., $\lim_{t \to 0+} T(t)x = x$. Then we know [5] that $\lim_{t \to 0+} S_\gamma(t)x = x$ so that the mapping $t \mapsto S_\gamma(t)x$, $t \ge 0$ is continuous and tends to zero as $t \to +\infty$. Let $p, q \in (1, \infty)$, and let $q(\gamma\beta - 1) > -1$. Assume that $p(x) \equiv r_2(x) \equiv p$ ($x \in [0,1]$) and $r_1 \in \mathcal{P}([0,1])$. Then $\|R(\cdot)\|_{L^q[0,1]} < \infty$ and the computation similar to that established in [20, Remark 2.14(ii)] shows that $m_t < \infty$ for all $t \ge 0$ as well as that the mapping $t \mapsto m_t$, $t \ge 0$ is continuous and satisfies $m_t \le \text{Const.} t^{\nu(-1-\gamma)}$, $t \ge 1$, where $\nu \in (0,1)$ is chosen so that $(1-\nu)(1+\gamma) > 1$. By Lemma 3.1, we get $\hat{m}. \in C_0([0,\infty) : L^{r_2(x)}([0,1] : X))$. Writing the first integral in (i) of Proposition 5.2 as $\int_0^{t+x} R(t+x-s)q(s)\, ds = \int_0^1 R(s)q(t+x-s)\, ds + \int_1^{t+x} R(s)q(t+x-s)\, ds$ ($t \ge 0$, $x \in [0,1]$), and using the growth order of $\|R(\cdot)\|$, it can be simply shown that the validity of condition

$$\lim_{t \to +\infty} \left\| \|q(t+x-\cdot)\|_{L^p[0,1]} + \int_0^{t+x-1} \frac{\|q(s)\|}{1+s^\gamma}\, ds \right\|_{L^{r_1(x)}[0,1]} = 0$$

yields that, for every $t \ge 0$, the mapping $x \mapsto \int_0^{t+x} R(t+x-s)q(s)\, ds$, $x \in [0,1]$ belongs to the space $L^{r_1(x)}([0,1] : X)$ and (12) holds. Therefore, Proposition 5.2 is applicable.

ACKNOWLEDGMENTS

The authors express their sincere gratitude to Mrs. Nadya Columbus who invited them to add a contribution to this edited book.

REFERENCES

[1] Diagana, T.–Kostić, M.: *Generalized almost periodic and generalized asymptotically almost periodic type functions in Lebesgue spaces with variable exponents $L^{p(x)}$*, Filomat, accepted.

[2] Diagana, T.–Zitane, M.: *Stepanov-like pseudo-almost automorphic functions in Lebesgue spaces with variable exponents $L^{p(x)}$*, Electron. J. Differential Equations 2013, no. **188**, 20 pp.

[3] Ding, H.-S.—Liang, J.— Xiao, T.-J.: Almost automorphic solutions to nonautonomous semilinear evolution equations in Banach spaces, *Nonlinear Anal.*, **73** (2010), 1426–1438.

[4] Fan, Z.—Liang, J.—Xiao, T.-J.: On Stepanov-like (pseudo)-almost automorphic functions, *Nonlinear Anal.*, **74** (2011), 2853–2861.

[5] Kostić, M.: *Almost Periodic and Almost Automorphic Solutions to Integro-Differential Equations,* W. de Gruyter, Berlin, 2019.

[6] Kostić, M.: Generalized almost automorphic and generalized asymptotically almost automorphic solutions of abstract Volterra integro-differential inclusions, *Fractional Diff. Calc.* **8** (2018), 255–284.

[7] Bazhlekova, E.: *Fractional evolution equations in Banach spaces,* PhD Thesis, Eindhoven University of Technology, Eindhoven, 2001.

[8] Diethelm, K.: *The Analysis of Fractional Differential Equations: An Application-Oriented Exposition Using Differential Operators of Caputo Type,* Springer-Verlag, Berlin, 2010.

[9] Kilbas, A. A.—Srivastava, H. M.—Trujillo, J. J.: *Theory and Applications of Fractional Differential Equations,* Elsevier Science B.V., Amsterdam, 2006.

[10] Kostić, M.: *Abstract Volterra Integro-Differential Equations,* Taylor and Francis Group/CRC Press, Boca Raton, Fl., 2015.

[11] Diening, L.—Harjulehto, P.—Hästüso, P.—Ruzicka, M.: Lebesgue and Sobolev Spaces with Variable Exponents, *Lecture Notes in Mathematics,* 2011. Springer, Heidelberg, 2011.

[12] Fan, X. L.–Zhao, D.: On the spaces $L^{p(x)}(O)$ and $W^{m,p(x)}(O)$, *J. Math. Anal. Appl.* **263** (2001), 424–446.

[13] Diagana, T.: *Almost Automorphic Type and Almost Periodic Type Functions in Abstract Spaces,* Springer, New York, 2013.

[14] Nguyen, P. Q. H.: On variable Lebesgue spaces, Thesis PhD, Kansas State University. Pro- Quest LLC, Ann Arbor, MI, 2011. 63 pp.

[15] Diagana, T.–Zitane, M.: Weighted Stepanov-like pseudo-almost periodic functions in Lebesgue space with variable exponents $L^{p(x)}$, *Afr. Diaspora J. Math.* **15** (2013), 56–75.

[16] Guérékata, G. M. N'.–Pankov, A.: Stepanov-like almost automorphic functions and monotone evolution equations, *Nonlinear Anal.* **68** (2008), 2658–2667.

[17] Bohr, H.–Følner, E.: On some types of functional spaces: A contribution to the theory of almost periodic functions, *Acta Math.* **76** (1944), 31–155.

[18] Levitan, M.-Zhikov, V. V.: *Almost Periodic Functions and Differential Equations,* Cambridge Univ. Press, London, 1982.

[19] Kováčik, O.–Rákosník, J.: On spaces $L^{p(x)}$ and $W^{k,p(x)}$, *Czechoslovak Math. J.* **41** (1991), 592–618.

[20] Kostić, M.: Existence of generalized almost periodic and asymptotic almost periodic solutions to abstract Volterra integro-differential equations, *Electron. J. Differential Equations*, vol. 2017, no. **239** (2017), 1–30.

[21] Guérékata, G.M.N'.–Kostić, M.: Generalized asymptotically almost periodic and generalized asymptotically almost automorphic solutions of abstract multiterm fractional differential inclusions, *Abstract Appl. Anal.*, volume 2018, Article ID 5947393, 17 pages https://doi.org/10.1155/2018/5947393.

[22] Guérékata, G.M.N'.: *Almost Automorphic and Almost Periodic Functions in Abstract Spaces,* Kluwer Acad. Publ., Dordrecht, 2001.

In: Recent Studies in Differential Equations
Editor: Henry Forster

ISBN: 978-1-53618-389-4
© 2020 Nova Science Publishers, Inc.

Chapter 2

EXISTENCE AND REGULARITY OF SOLUTIONS FOR SOME NONLINEAR SECOND ORDER DIFFERENTIAL EQUATION IN BANACH SPACES

Issa Zabsonré[*] *and Micailou Napo*
Université Joseph KI-ZERBO,
Unité de Recherche et de Formation en Sciences Exactes et Appliquées,
Département de Mathématiques, Ouagadougou, Burkina Faso

Abstract

In this work, we study the existence and regularity of solutions for some nonlinear second order differential equation. The delayed part is assumed to be locally lipschitz. Firstly, we show the existence of the mild solutions. Secondly, we give sufficiently conditions ensuring the existence of strict solutions.

Keywords: cosine family, mild and strict solutions, second order partial differential

Mathematics Subject Classification: 47D09, 34G10, 60J65

---
[*]Corresponding Author's Email: zabsonreissa@yahoo.fr.

1. INTRODUCTION

In this work, we study the existence and regularity of solutions for the following partial functional equation

$$\begin{cases} u''(t) = Au(t) + f(t, u_t, u'_t) \text{ for } t \geq 0 \\ u_0 = \varphi \in \mathcal{C} = C^1([-r, 0]; X) \\ u'_0 = \varphi' \in \mathcal{C}, \end{cases} \quad (1.1)$$

where A is the (possibly unbounded) infinitesimal generator of a strongly continuous cosine family of linear operators in X, $\mathcal{C}$ denotes the space of continuous differentiable functions from $[-r, 0]$ to X endowed with the following norm $\|h\| = |h| + |h'|$ for all $h \in \mathcal{C}$, for every $t \geq 0$, u_t denotes the history function of $\mathcal{C}$ defined by

$$u_t(\theta) = u(t + \theta) \text{ for } -r \leq \theta \leq 0,$$

$f : \mathbb{R}^+ \times \mathcal{C} \times \mathcal{C} \to X$ is a given function.

In [1] the authors study some semi-linear second order initial value problem. They also unify and simplify some ideas from the theory of strongly continuous cosine families of linear operators in Banach spaces. In [2], the authors reveal three properties of cosine families, distinguishing them from semigroups of operators.

In [3], by using the theory of strongly continuous cosine families of linear operators in Banach, the author investigated the existence of solutions of the following semilinear second order differential initial value problem

$$\begin{cases} u''(t) = Au(t) + g(t, u(t), u'(t)) \text{ for } t \in [0, T] \\ u(0) = u_0 \in X \quad u'(0) = u_1 \in X. \end{cases} \quad (1.2)$$

Using the theory of strongly continuous cosine families of linear operators in Banach space, we prove in this paper the existence of the mild and strict solutions. The organization of this work is as follows, in section 2, we collect some background materials required throughout the paper. In section 3, we study the existence of local mild solutions of equation (1.1) and we show the global continuation of solutions. We prove that in the case of local existence, the solutions blows up, we also show the dependence continuous with the initial data. In section 4, we will show the existence of strict solutions for equation (1.1). For illustration, we propose to study the existence of solutions for some partial functional equations with diffusion.

2. PRELIMINARY RESULTS

Definition 2.1. *A one parameter family $C(t)$, $t \in \mathbb{R}$ of bounded linear operators mapping the Banach space X into itself is called a strongly continuous cosine family if and only if*
i) $C(s+t) + C(s-t) = 2C(s)C(t)$ *for all* $s, t \in \mathbb{R}$,
ii) $C(0) = I$,
iii) $C(t)x$ *is continuous in t on $\mathbb{R}$ for each fixed $x \in X$.*

If $C(t), t \in \mathbb{R}$ is a strongly continuous cosine family in X, then $S(t)$ defined by

$$S(t)x = \int_0^t C(t)x \, ds \text{ for } x \in X, \ t \in \mathbb{R}. \tag{2.1}$$

is a one parameter family of operators in X.

Lemma 2.2. *[1] Let $C(t)$, $t \in \mathbb{R}$ be a strongly continuous cosine family in X. The following are true:*
i) $C(t) = C(-t)$ *for all $t \in \mathbb{R}$,*
ii) $C(s), S(s), C(t)$, *and* $S(t)$ *commute for all* $s, t \in \mathbb{R}$,
iii) $S(t)x$ *is continuous in t on $\mathbb{R}$ for each fixed $x \in X$,*
iv) $S(s+t) + S(s-t) = 2S(s)C(t)$ *for all* $s, t \in \mathbb{R}$,
v) $S(s+t) = S(s)C(t) + S(t)C(s)$ *for all* $s, t \in \mathbb{R}$,
vi) $S(t) = -S(-t)$ *for all $t \in \mathbb{R}$.*

Definition 2.3. *The infinitesimal generator of a strongly continuous cosine family $C(t)$, $t \in \mathbb{R}$ is the operator $A : X \to X$ defined by*

$$Ax = \frac{d^2 C(t)x}{dt^2}\bigg|_{t=0}.$$

$D(A) = \{x \in X : C(t)x \text{ is a twice continuously differentiable function of } t\}$.

We shall also make use of the set

$$E = \{x : C(t)x \text{ is a once continuously differentiable function of } t\}.$$

Lemma 2.4. *[1] Let $C(t)$, $t \in \mathbb{R}$, be a strongly continuous cosine family in X with infinitesimal generator A. The following are true.*

i) $D(A)$ is dense in X and A is a closed operator in X

ii) if $x \in X$ and $r, s \in \mathbb{R}$, then $z = \int_r^s S(u)x\,du \in D(A)$ and $Az = C(s)x - C(r)x$,

iii) If $x \in X$ and $r, s \in \mathbb{R}$, then $z = \int_0^s \int_0^r C(u)C(v)x\,du\,dv \in D(A)$ and $Az = \dfrac{1}{2}(C(r+s)x - C(s-r)x)$

iv) if $x \in X$, then $S(t)x \in E$,

v) if $x \in X$, then $S(t)x \in D(A)$ and $\dfrac{dC(t)x}{dt} = AS(t)x$,

vi) if $x \in D(A)$, then $C(t)x \in D(A)$ and $\dfrac{d^2C(t)x}{dt^2} = AC(t)x = C(t)Ax$,

vii) if $x \in E$, then $\lim_{t \to 0} AS(t)x = 0$,

viii) if $x \in E$, then $S(t)x \in D(A)$ and $\dfrac{d^2S(t)x}{dt^2} = AS(t)x$,

ix) if $x \in D(A)$, then $S(t)x \in D(A)$ and $AS(t)x = S(t)Ax$,

x) $C(t+s) - C(t-s) = 2AS(t)S(s)$ for all $s, t \in \mathbb{R}$.

Throughout this chapter, we assume that

(H_0) A is the infinitesimal generator of a strongly continuous cosine family of linear operators on a Banach space X.

By Proposition 2.4, **(H_0)** implies that the operator A is densely defined in X, i.e., $\overline{D(A)} = X$. We have the following result.

Lemma 2.5. *[1] Assume that **(H_0)** holds. Then there are constants $M \geq 1$ and $\omega \geq 0$ such that*

$$\|C(t)\| \leq Me^{\omega|t|} \text{ and } \|S(t)\| \leq Me^{\omega|t|} \text{ and for } t \in \mathbb{R}.$$

Theorem 2.6. *[3] If $g : [0, T] \times X \times X \to X$ is continuous and u is a solution of equation (1.2), then u is a solution of the integral equation*

$$u(t) = C(t)u_0 + S(t)u_1 + \int_0^t S(t-s)g(s, u(s), u'(s))\,ds \text{ for } t \geq 0,$$

3. LOCAL EXISTENCE, GLOBAL CONTINUATION AND BLOWING UP OF SOLUTIONS

Lemma 3.1. *Assume that* (H_0) *holds. If u is a solution of equation* (1.1), *then*

$$u(t) = C(t)\varphi(0) + S(t)\varphi'(0) + \int_0^t S(t-s)f(s, u_s, u'_s)ds \text{ for } t \geq 0, \quad (3.1)$$

Proof. It is just a consequence of Theorem 2.6. In fact, let us pose $g(t, u(t), u'(t)) = f(t, u_t, u'_t)$ for $t \geq 0$. Then we get the desired result. ∎

Remark: The converse is not true. In fact if u satisfies equation (5.1), u may be not twice continuously differentiable, that is why we distinguish between mild and strict solutions.

Definition 3.2. *We say that a continuous function* $u : [-r, +\infty[\to X$ *is a strict solution of equation* (1.1) *if the following conditions hold*
(i) $x \in C^1([0, +\infty[; X) \cap C^2((0, +\infty[; X)$.
(ii) x *satisfies equation* (1.1) *on* $[0, +\infty[$.
(iii) $u(\theta) = \varphi(\theta)$ *for* $-r \leq \theta \leq 0$.

Definition 3.3. *We say that a continuous function* $u : [-r, +\infty[\to X$ *is a mild solution of equation* (1.1) *if u satisfies the following equation*

$$\begin{cases} u(t) = C(t)\varphi(0) + S(t)\varphi'(0) + \int_0^t S(t-s)f(s, u_s, u'_s)ds \text{ for } t \geq 0 \\ u_0 = \varphi(\theta) \text{ for } -r \leq \theta \leq 0. \\ u'_0 = \varphi'(\theta) \text{ for } -r \leq \theta \leq 0. \end{cases}$$

In the following, we give a local existence of mild solutions of equation (1.1). For this purpose, we make the following assumption.

(H_1) f is locally lipschitz, that is, for each $\alpha > 0$ there is a constant $c_0(\alpha) > 0$ such that if $\varphi_1, \varphi_2, \varphi_3, \varphi_4 \in \mathcal{C}$ $\|\varphi_1\|, \|\varphi_2\|, \|\varphi_3\|, \|\varphi_4\| \leq \alpha$ then

$$|f(t, \varphi_1, \varphi_2) - f(t, \varphi_3, \varphi_4)| \leq c_0(\alpha)(|\varphi_1 - \varphi_3| + |\varphi_2 - \varphi_4|).$$

Theorem 3.4. *Assume that* (H_0) *and* (H_1) *hold. Let* $\varphi \in \mathcal{C}$ *such that* $\varphi(0) \in D(A)$ *and* $\varphi'(0) \in E$. *Then, there exists a maximal interval of existence* $[-r, b_\varphi[$ *and a unique mild solution* $u(., \varphi)$ *of equation* (1.1) *defined on* $[-r, b_\varphi[$ *and either*

$$b_\varphi = +\infty \text{ or } \overline{\lim}_{t \to b_\varphi^-}(|u(t, \varphi)| + |u'(t, \varphi)|) = +\infty.$$

Moreover, $u(t, \varphi)$ *is a continuous function of* φ *in the sense that if* $\varphi \in \mathcal{C}$ *and* $t \in [0, b_\varphi[$, *then there exist positive constants* k *and* ε *such that, for* $\varphi, \psi \in \mathcal{C}$ *and* $\|\varphi - \psi\| < \varepsilon$, *we have*

$t \in [0, b_\psi[$ *and* $|u(s, \varphi) - u(s, \psi)| + |u'(s, \varphi) - u,(s, \psi)| \le k\|\varphi - \psi\|$ *for all* $s \in [-r, t]$.

Proof. Let $b_1 > 0$. The local lipschitz condition on f implies that for each $\alpha > 0$, there exists $c_0(\alpha)$ such that for $\varphi \in \mathcal{C}$ with $\|\varphi\| < \alpha$ we have

$$|f(t, \varphi, \varphi')| \le c_0(\alpha)\|\varphi\| + |f(t, 0, 0)| \le c_0(\alpha)\alpha + \sup_{s \in [0, b_1]} |f(s, 0, 0)|,$$

with for given $\varphi \in \mathcal{C}$, $\alpha = \|\varphi\| + 1$ and $c_1(\alpha) = c_0(\alpha)\alpha + \sup_{s \in [0, b_1]} |f(s, 0, 0)|$.

Consider the following set

$$Z_\varphi = \left\{ \begin{array}{l} u \in C^1([-r, b_1]; X) : u(s) = \varphi(s), \; u'(s) = \varphi'(s) \text{ if } s \in [-r, 0] \\ \text{and } \sup_{s \in [0, b_1]}(|u(s) - \varphi(0)| + |u'(s) - \varphi'(0)|) \le 1, \end{array} \right\}$$

then Z_φ is a closed set of $C^1([-r, b_1]; X)]$. Consider the mapping

$$\mathcal{K} : Z_\varphi \to C^1([-r, b_1]; X)$$

defined by

$$\begin{cases} \mathcal{K}(u)(t) = C(t)\varphi(0) + S(t)\varphi'(0) + \int_0^t S(t-s)f(s, u_s, u'_s)ds \text{ for } t \ge 0 \\ \mathcal{K}(u_0)(t) = \varphi(t) \text{ for } t \in [-r, 0] \\ (\mathcal{K}(u_0))'(t) = \varphi'(t) \text{ for } t \in [-r, 0]. \end{cases}$$

We will show that $\mathcal{K}(Z_\varphi) \subset Z_\varphi$. Let $u \in Z_\varphi$ and $t \in [0, b_1]$, using Proposition 2.5, we have

$$|\mathcal{K}(u)(t) - \varphi(0)| \leq |C(t)\varphi(0) - \varphi(0)| + |S(t)\varphi'(0)| + \left|\int_0^t S(t-s)f(s, u_s, u'_s)ds\right|$$

$$\leq |C(t)\varphi(0) - \varphi(0)| + |S(t)\varphi'(0)| + Me^{\omega t}\int_0^t e^{-\omega s}|f(s, u_s, u'_s)|ds$$

$$\leq |C(t)\varphi(0) - \varphi(0)| + |S(t)\varphi'(0)| + Me^{\omega t}\int_0^t |f(s, u_s, u'_s)|ds.$$

Since

$$|u(s)| - |\varphi(0)| + |u'(s)| - |\varphi'(0)| \leq |u(s) - \varphi(0)| + |u'(s) - \varphi'(0)|$$
$$\leq 1 \quad \text{for } s \in [0, b_1]$$

and $\alpha = \|\varphi\| + 1$, we deduce that $\|u_s\| \leq 1 + \|\varphi\| = \alpha$ for $s \in [0, b_1]$. Then

$$|f(s, u_s, u'_s)| \leq c_0(\alpha)\|u_s\| + |f(s, 0, 0)| \leq c_1(\alpha). \tag{3.2}$$

If we choose b_1 sufficiently small such that

$$\sup_{s \in [0, b_1]} \left\{|C(s)\varphi(0) - \varphi(0)| + |S(s)\varphi'(0)| + Me^{\omega s}c_1(\alpha)s\right\} < \frac{1}{2},$$

consequently

$$|\mathcal{K}(u)(t) - \varphi(0)| \leq |C(t)\varphi(0) - \varphi(0)| + |S(t)\varphi'(0)| + Me^{\omega t}c_1(\alpha)t \text{ for } t \in [0, b_1].$$
$$< \frac{1}{2},$$

On the other hand using equation (2.1), Proposition 2.4 and Proposition 2.5, we have

$$(\mathcal{K}(u))'(t) = C'(t)\varphi(0) + S'(t)\varphi'(0) + \int_0^t C(t-s)f(s, u_s, u'_s)ds \text{ for } t \geq 0$$

$$|(\mathcal{K}(u))'(t) - \varphi'(0)| \leq |AS(t)\varphi(0)| + |C(t)\varphi'(0) - \varphi'(0)| + \left|\int_0^t C(t-s)f(s, u_s, u'_s)ds\right|$$

$$\leq |AS(t)\varphi(0)| + |C(t)\varphi'(0) - \varphi'(0)| + Me^{\omega t}\int_0^t e^{-\omega s}|f(s, u_s, u'_s)|ds$$

$$\leq |AS(t)\varphi(0)| + |C(t)\varphi'(0) - \varphi'(0)| + Me^{\omega t}\int_0^t |f(s, u_s, u'_s)|ds.$$

From [1] (in Proposition 2.4), $t \to C(t)\varphi(0) + S(t)\varphi'(0)$ belongs to $C^2([0, b_1]; X)$, we also choose b_1 sufficiently small such that

$$\sup_{s\in[0,b_1]} \left\{ |AS(s)\varphi(0)| + |C(s)\varphi'(0) - \varphi'(0)| + Me^{\omega s}c_1(\alpha)s \right\} < \frac{1}{2},$$

consequently

$$|(\mathcal{K}(u))'(t) - \varphi'(0)| \leq |AS(t)\varphi(0)| + |C(t)\varphi'(0) - \varphi'(0)| + Me^{\omega t}c_1(\alpha)t \text{ for } t \in [0, b_1].$$
$$< \frac{1}{2},$$

Finally we have

$$|\mathcal{K}(u)(t) - \varphi(0)| + |(\mathcal{K}(u))'(t) - \varphi'(0)| < 1,$$

hence $\mathcal{K}(Z_\varphi) \subset Z_\varphi$.

Let $u, v \in Z_\varphi$ and $t \in [0, b_1]$. Then using Proposition 2.5

$$\begin{aligned}
|\mathcal{K}(u)(t) - \mathcal{K}(v)(t)| &= \left| \int_0^t S(t-s)(f(s, u_s, u'_s) - f(s, v_s, v'_s))ds \right| \\
&\leq Me^{\omega t} \int_0^t |f(s, u_s, u'_s) - f(s, v_s, v'_s)|ds \\
&\leq Me^{\omega b_1} c_0(\alpha) \int_0^t \|u_s - v_s\| ds \\
&\leq Me^{\omega b_1} c_0(\alpha) b_1 \|u - v\|.
\end{aligned}$$

It follows that

$$\begin{aligned}
Me^{\omega b_1} c_0(\alpha) b_1 &\leq Me^{\omega b_1} c_1(\alpha) b_1 \\
&\leq \sup_{s\in[0,b_1]} \left\{ |C(s)\varphi(0) - \varphi(0)| + |S(s)\varphi'(0)| + Me^{\omega s} c_1 s \right\} \\
&< \frac{1}{2},
\end{aligned}$$

which shows that

$$|\mathcal{K}(u)(t) - \mathcal{K}(v)(t)| < \frac{1}{2}\|u - v\|.$$

Using the same reasoning, we have

$$|(\mathcal{K}(u))'(t) - (\mathcal{K}(v))'(t)| = \left| \int_0^t C(t-s)(f(s, u_s, u'_s) - f(s, v_s, v'_s))ds \right|$$

$$< \frac{1}{2}\|u - v\|.$$

Adding the two previous equations it follows that $\mathcal{K}$ is a strict contraction in Z_φ. Thus by a fixed point theorem, $\mathcal{K}$ has a unique fixed point u in Z_φ. We conclude that equation (1.1) has one and only one mild solution which is defined on $[-r, b_1]$ and denoted by $u(., \varphi)$. Using the same arguments, we can show that $u(., \varphi)$ can be extended to a maximal interval of existence $[0, b_\varphi[$. If we assume that $b_\varphi < +\infty$ and $\overline{\lim}_{t \to b_\varphi^-}(|u(t, \varphi)| + |u'(t, \varphi)|) < +\infty$, then there exists a constant $\alpha > 0$ such that $(|u(t, \varphi)| + |u'(t, \varphi)|) \leq \alpha$ for all $t \in [0, b_\varphi[$. We claim that $u(., \varphi)$ and $u'(., \varphi)$ are uniformly continuous. Consequently

$$\lim_{t \to b_\varphi^-} (u(t, \varphi) + u'(t, \varphi)) \text{ exists,}$$

which contradicts the maximality of $[0, b_\varphi[$. Let us show the uniform continuity of $u(., \varphi)$ and $u'(., \varphi)$. Let t, $t+h \in [0, b_\varphi[$, $h > 0$ and $\theta \in [-r, 0]$. If $t+\theta \geq 0$, then the map $t \mapsto C(t+\theta)\varphi(0) + S(t+\theta)\varphi'(0)$ is uniformly continuous. On the other hand let us pose $u(., \varphi) = u$, we have

$$u(t+h+\theta) - u(t+\theta) = C(t+h+\theta)\varphi(0) - C(t+\theta)\varphi(0) + S(t+h+\theta)\varphi'(0) - S(t+\theta)\varphi'(0)$$

$$+ \int_0^{t+\theta+h} S(t+\theta+h-s)f(s, u_s, u'_s)ds - \int_0^{t+\theta} S(t+\theta-s)f(s, u_s, u'_s)ds$$

$$= C(t+h+\theta)\varphi(0) - C(t+\theta)\varphi(0) + S(t+h+\theta)\varphi'(0) - S(t+\theta)\varphi'(0)$$

$$+ \int_0^{t+\theta} S(s)f(t+\theta+h-s, u_{t+\theta+h-s}, u'_{t+\theta+h-s})ds$$

$$+ \int_{t+\theta}^{t+\theta+h} S(s)f(t+\theta+h-s, u_{t+\theta+h-s}, u'_{t+\theta+h-s})ds$$

$$- \int_0^{t+\theta} S(s)f(t+\theta-s, u_{t+\theta-s}, u'_{t+\theta-s})ds$$

$$= C(t+h+\theta)\varphi(0) - C(t+\theta)\varphi(0) + S(t+h+\theta)\varphi'(0) - S(t+\theta)\varphi'(0)$$

$$+ \int_0^{t+\theta} S(s)\Big[f(t+\theta+h-s, u_{t+\theta+h-s}, u'_{t+\theta+h-s}) - f(t+\theta-s, u_{t+\theta-s}, u'_{t+\theta-s})\Big]ds$$

$$+ \int_{t+\theta}^{t+\theta+h} S(s)f(t+\theta+h-s, u_{t+\theta+h-s}, u'_{t+\theta+h-s})ds.$$

Thus, using equation (3.2) and the local Lipschitz condition of f, we have

$$|u(t+h+\theta,\varphi) - u(t+\theta,\varphi)|$$

$$\leq |C(t+h+\theta)\varphi(0) - C(t+\theta)\varphi(0)| + |S(t+h+\theta)\varphi'(0) - S(t+\theta)\varphi'(0)|$$

$$+ Me^{\omega b_\varphi} c_1(\alpha) h + Me^{\omega b_\varphi} c_0(\alpha) \int_0^t (|u_{s+h}(.) - u_s(.)| + |u'_{s+h}(.) - u'_s(.,)|) ds.$$

If $t + \theta < 0$. Let $h_0 > 0$ sufficiently small such for $h \in]0, h_0[$

$$|u_{t+h}(\theta) - u_t(\theta)| \leq \sup_{-r \leq \sigma \leq 0} |u(\sigma+h) - u(\sigma)| = |u_h - \varphi|_\infty$$

Since the map $t \mapsto C(t)\varphi(0) + S(t)\varphi'(0)$ is uniformly continuous, consequently, for $t, t + h \in [0, b_\varphi[$ and $h \in]0, h_0[$, we have

$$|u_{t+h}(.) - u_t(.)|_\infty \leq \delta_1(h) + \delta_2(h) + c_1(\alpha) Me^{\omega b_\varphi} h$$

$$+ Me^{\omega b_\varphi} c_0(\alpha) \int_0^t \|u_{s+h}(.) - u_s(.)\| ds$$

where

$\delta_1(h) = \|u_h - \varphi\|$ and $\delta_2(h) = \sup\limits_{\substack{t \\ t+h \in [0, b_\varphi[}} \Big(|C(t+h)\varphi(0) - C(t)\varphi(0)| + |S(t+h)\varphi'(0) - S(t)\varphi'(0)|\Big).$

From [1] (in Proposition 2.4), $t \to C(t)\varphi(0) + S(t)\varphi'(0)$ belongs to $C^2([0, b_\varphi]; X)$, by a similar reasoning, we also have

$$|u'_{t+h}(.) - u'_t(.)|_\infty \leq \delta'_1(h) + \delta'_2(h) + c_1(\alpha) Me^{\omega b_\varphi} h$$

$$+ Me^{\omega b_\varphi} c_0(\alpha) \int_0^t \|u_{s+h}(.) - u_s(.)\| ds$$

where

$\delta'_1(h) = \|u'_h - \varphi\|$ and $\delta'_2(h) = \sup\limits_{\substack{t \\ t+h \in [0, b_\varphi[}} \Big(|AS(t+h)\varphi(0) - AS(t)\varphi(0)| + |C(t+h)\varphi'(0) - C(t)\varphi'(0)|\Big).$

By Gronwall's lemma, it follows that

$$\|u_{t+h}(., \varphi) - u_t(., \varphi)\| \leq \gamma(h) \exp[2c_0(\alpha) Me^{\omega b_\varphi} b_\varphi],$$

with
$$\gamma(h) = \delta_1(h) + \delta_2(h) + \delta'_1(h) + \delta'_2(h) + 2c_1(\alpha)Me^{\omega b_\varphi}h.$$
This completes that u and u' are uniformly continuous and u can be extended over $[0, b_\varphi + \eta]$, which contradicts the maximality of $[0, b_\varphi[$. Using the same reasoning, one can show a similar result for $h < 0$.

Now, we want to prove that the solution depends continuously on initial data. Let $\varphi \in C$ and $t \in [0, b_\varphi[$ be fixed. Set
$$\alpha = 1 + \sup_{-r \leq s \leq t} \|u_s(., \varphi)\|$$
and
$$c(t) = Me^{\omega t}\exp(Me^{\omega t}c_0(\alpha)t).$$
Let $\varepsilon \in]0, 1[$ and $\psi \in C$ such that $\|\varphi - \psi\| < \varepsilon$. Then
$$\|\psi\| \leq \|\varphi\| + \varepsilon < \alpha.$$
We define
$$b_0 := \sup\{s > 0 : \|u_\sigma(., \psi)\| \leq \alpha \text{ for } \sigma \in [0, s]\}.$$
If we suppose that $b_0 < t$, we obtain for $s \in [0, b_0]$
$$|u_s(., \varphi) - u_s(., \psi)| \leq Me^{\omega t}\|\varphi - \psi\|$$
$$+ Me^{\omega t}c_0(\alpha)\int_0^s |u_\sigma(., \varphi) - u_\sigma(., \psi)|ds.$$
From [1] (in Proposition 2.4), $t \to C(t)\varphi(0) + S(t)\varphi'(0)$ belongs to $C^2([0, b_0]; X)$, then $\|S(t)A\|$ is bounded uniformly in finite intervals of t, thus we also have
$$|u'_s(., \varphi) - u'_s(., \psi)| \leq Me^{\omega t}\|\varphi - \psi\|$$
$$+ Me^{\omega t}c_0(\alpha)\int_0^s |u_\sigma(., \varphi) - u_\sigma(., \psi)|ds.$$
By Gronwall's lemma, we deduce that
$$\|u_s(., \varphi) - u_s(., \psi)\| \leq c(t)\|\varphi - \psi\|. \tag{3.3}$$

This implies that

$$\|u_s(.,\psi)\| \le c(t)\varepsilon + \alpha - 1 < \alpha \text{ for all } s \in [0, b_0].$$

It follows that b_0 cannot be the largest number $s > 0$ such that $\|u_s(.,\psi)\| < \alpha$, for $\sigma \in [0, s]$. Thus $b_0 \ge t$ and $t < b_\psi$. Furthermore, $\|u_s(.,\varphi)\| < \alpha$ for $s \in [0, t]$, then using the inequality (3.3) we deduce the continuous dependence on the initial data. ∎

Corollary 3.5. *Assume that* (H_0) *and* (H_1) *hold. Let* $\varphi \in \mathcal{C}$ *such that* $\varphi(0) \in D(A)$ *and* $\varphi'(0) \in E$. *Let* k_1 *be a continuous function on* $\mathbb{R}^+$ *and* $k_2 \in L^1_{loc}(\mathbb{R}^+; \mathbb{R}^+)$ *be such that* $|f(t, \varphi, \varphi')| \le k_1(t)\|\varphi\| + k_2(t)$ *for* $t \ge 0$ *and* $\varphi, \varphi' \in \mathcal{C}$. *Then equation* (1.1) *has a unique mild solution which is defined for all* $t \ge 0$.

Proof. Let $[-r, b_\varphi[$ denote the maximal interval of existence of the mild solution $u(t, \varphi)$ of equation (1.1). Then

$$b_\varphi = +\infty \text{ or } \overline{\lim}_{t \to b_\varphi^-}(|u(t, \varphi)| + |u'(t, \varphi)|) = +\infty.$$

If $b_\varphi < +\infty$, then $\overline{\lim}_{t \to b_\varphi^-}(|u(t, \varphi)| + |u'(t, \varphi)|) = +\infty$. Since $C(t), S(t) \in \mathcal{B}(X)$ for $t \ge 0$, then there exists $M > 0$ such that $|C(t)|, |S(t)| \le M$ for all $t \in [0, b_\varphi]$. Thus, we have

$$|u(t, \varphi)| \le |C(t)\varphi(0)| + |S(t)\varphi'(0)| + \int_0^t |S(t-s)| |f(s, u_s, u'_s)| ds$$

$$\le k_0 + \int_0^t Mk_1(s)\|u_s\| ds \text{ for } t \in [0, b_\varphi[,$$

where $k_0 = M(|\varphi| + |\varphi'|) + \int_0^{b_\varphi} Mk_2(s)ds$. Using Proposition 2.5 and since $\|S(t)A\|$ is bounded uniformly in finite intervals of t, then

$$|u'(t, \varphi)| \le |S(t)A\varphi(0)| + |C(t)\varphi'(0)| + \int_0^t |C(t-s)| |f(s, u_s, u'_s)| ds$$

$$\le k_0 + \int_0^t Mk_1(s)\|u_s\| ds \text{ for } t \in [0, b_\varphi[,$$

. By Gronwall's lemma, we deduce that

$$\|u_t(\varphi)\| \leq 2k_0 \exp\left(2M \int_0^t k_1(s)\right)ds < +\infty \text{ for } t \in [0, b_\varphi[,$$

and
$$\overline{\lim}_{t \to b_\varphi^-}(|u(t,\varphi)| + |u'(t,\varphi)|) < +\infty,$$

which gives a contradiction. ∎

As an immediat consequence, we get the following result.

Corollary 3.6. *Assume that* (H_0) *and* (H_1) *hold and there exists a positive constant L such that for* $\varphi_1, \varphi_2 \in \mathcal{C}$

$$|f(t, \varphi_1, \varphi_1') - f(t, \varphi_2, \varphi_2')| \leq L\|\varphi_1 - \varphi_2\| \text{ for } t \geq 0.$$

Let $\varphi \in \mathcal{C}$ *such that* $\varphi(0) \in D(A)$ *and* $\varphi'(0) \in E$. *Then equation* (1.1) *has a unique mild solution which is defined for all* $t \geq 0$.

In the sequel, we give some estimations of solutions.

Lemma 3.7. *Assume that* (H_0) *and* (H_1) *hold and and there exists a positive constant L such that for* $\varphi_1, \varphi_2 \in \mathcal{C}$

$$|f(t, \varphi_1, \varphi_1') - f(t, \varphi_2, \varphi_2')| \leq L\|\varphi_1 - \varphi_2\| \text{ for } t \geq 0.$$

Let u *and* $\widehat{u}$ *be the mild solutions of equation* (1.1) *corresponding respectively to*
φ *and* $\widehat{\varphi} \in \mathcal{C}$. *Then*

$$\|u_t - \widehat{u}_t\| \leq 2M\|\varphi - \widehat{\varphi}\|e^{(\omega+2ML)t}.$$

Proof. Let u and $\widehat{u}$ be the mild solutions of equation (1.1) corresponding respectively to
φ and $\widehat{\varphi} \in \mathcal{C}$, we have

$$|u(t) - \widehat{u}(t)| \leq Me^{\omega t}|\varphi(0) - \widehat{\varphi}(0)| + Me^{\omega t}|\varphi'(0) - \widehat{\varphi}'(0)| + ML\int_0^t e^{\omega(t-s)}\|u_s - \widehat{u}_s\|ds.$$

By using equation (2.1), Proposition 2.4 and since $\|S(t)A\|$ is bounded uniformly in finite intervals of t, then

$$|u'(t) - \widehat{u}'(t)| \leq Me^{\omega t}|\varphi(0) - \widehat{\varphi}(0)| + Me^{\omega t}|\varphi'(0) - \widehat{\varphi}'(0)| + ML\int_0^t e^{\omega(t-s)}\|u_s - \widehat{u}_s\|ds.$$

It follows that

$$|u(t) - \widehat{u}(t)| + |u'(t) - \widehat{u}'(t)| \leq 2Me^{\omega t}(|\varphi(0) - \widehat{\varphi}(0)| + |\varphi'(0) - \widehat{\varphi}'(0)|)$$

$$+ 2ML\int_0^t e^{\omega(t-s)}\|u_s - \widehat{u}_s\|ds.$$

By Gronwall's lemma, the result follows. ∎

4. EXISTENCE OF STRICT SOLUTIONS

Theorem 4.1. *Assume that* (H_0) *and* (H_1) *hold and* f *is continuously differentiable. Moreover we suppose that the partial derivatives* f_1, f_2 *and* f_3 *are locally lipschitz in classical sens, i.e there exists a positive constant* c_1 *such that for* $\varphi_1, \varphi_2 \in \mathcal{C}$

$$|f_i(t, \varphi_1, \varphi_1') - f_i(t, \varphi_2, \varphi_2')| \leq c_1\|\varphi_1 - \varphi_2\| \text{ for } t \geq 0, \ i = 1, 2, 3.$$

Let φ *be in* $C^3([-r, 0], X)$ *such that* $\varphi(0), \varphi''(0) \in D(A)$, $\varphi'(0), \varphi^{(3)}(0) \in E$, $\varphi''(0) = A\varphi(0) + f(0, \varphi, \varphi')$ *and* $\varphi^{(3)}(0) = A\varphi'(0)$. *Then the corresponding mild solution* u *is a strict solution of equation (1.1).*

Proof. Let φ be in $C^3([-r, 0], X)$ such that $\varphi(0), \varphi''(0) \in D(A)$, $\varphi'(0), \varphi^{(3)}(0) \in E$, $\varphi''(0) = A\varphi(0) + f(0, \varphi, \varphi')$ and $\varphi^{(3)}(0) = A\varphi'(0)$. Let u be the corresponding mild solution of equation (1.1) which is defined on some maximal interval $[0, b_\varphi[$ and let $a < b_\varphi$. Then by using the strict contraction principle, we can show that there exists a unique continuous function v such that

$$v(t) = \begin{cases} C(t)(A\varphi(0) + f(0, \varphi, \varphi')) + S(t)A\varphi'(0) \\ \quad + \int_0^t C(t-s)\Big[f_1(s, u_s, u_s') + f_2(s, u_s, u_s')u_s' + f_3(s, u_s, u_s')v_s\Big]ds \\ v_0 = \varphi''. \end{cases}$$

We introduce the function w defined by

$$\begin{cases} w(t) = \varphi'(0) + \int_0^t v(s)ds \text{ if } t \geq 0 \\ w(t) = \varphi'(t) \text{ if } -r \leq t \leq 0 \\ w'(t) = \varphi''(t) \text{ if } -r \leq t \leq 0. \end{cases}$$

We will show that $w = u'$. We can also see that

$$w_t = \varphi' + \int_0^t v_s ds \text{ for } t \in [0, a].$$

Consequently, the maps $t \to w_t$ and $t \to \int_0^t C(t-s)f(s, u_s, w_s)ds$ are continuously differentiable and the following formula holds

$$\frac{d}{dt} \int_0^t C(t-s)f(s, u_s, w_s)ds = \frac{d}{dt} \int_0^t C(s)f(t-s, u_{t-s}, w_{t-s})ds$$
$$= C(t)f(0, u_0, w_0) + \int_0^t C(t-s)\Big[f_1(s, u_s, w_s) + f_2(s, u_s, w_s)u'_s + f_3(s, u_s, w_s)v_s\Big]ds$$
$$= C(t)f(0, \varphi, \varphi') + \int_0^t C(t-s)\Big[f_1(s, w_s, w'_s) + f_2(s, u_s, w_s)u'_s + f_3(s, u_s, w_s)v_s\Big]ds,$$

which implies

$$\int_0^t C(s)f(0, \varphi, \varphi')ds = \int_0^t C(t-s)f(s, u_s, w_s)ds - \int_0^t \int_0^s C(s-\tau)\Big[f_1(\tau, u_\tau, w_\tau) + f_2(\tau, u_\tau, w_\tau)u'_\tau$$
$$+ f_3(\tau, u_\tau, w_\tau)v_\tau\Big]d\tau ds.$$

Consequently we have

$$w(t) = \varphi'(0) + \int_0^t C(s)A\varphi(0)\,ds + \int_0^t C(t-s)f(s, u_s, w_s)ds + \int_0^t S(s)A\varphi'(0)\,ds$$
$$- \int_0^t \int_0^s C(s-\tau)\Big[f_1(\tau, u_\tau, w_\tau) + f_2(\tau, u_\tau, w_\tau)u'_\tau + f_3(\tau, u_\tau, w_\tau)v_\tau\Big]d\tau ds$$
$$+ \int_0^t \int_0^s C(s-\tau)\Big[f_1(\tau, u_\tau, u'_\tau) + f_2(\tau, u_\tau, u'_\tau)u'_\tau + f_3(\tau, u_\tau, u'_\tau)v_\tau\Big]d\tau ds.$$

Since by equation (2.1) and Proposition 2.4, we have

$$\int_0^t C(s)A\varphi(0)ds = S(t)A\varphi(0)$$

$$\int_0^t S(s)A\varphi'(0)\,ds = C(t)\varphi'(0) - \varphi'(0),$$

it follows that

$$w(t) = S(t)A\varphi(0) + \int_0^t C(t-s)f(s, u_s, w_s)ds + C(t)\varphi'(0)$$
$$+ \int_0^t \int_0^s C(s-\tau)\Big[(f_1(\tau, u_\tau, u'_\tau) - f_1(\tau, u_\tau, w_\tau)) + (f_2(\tau, u_\tau, u'_\tau)u'_\tau - f_2(\tau, u_\tau, w_\tau)u'_\tau)$$
$$+ (f_3(\tau, u_\tau, u'_\tau)v_\tau) - f_3(\tau, u_\tau, w_\tau)v_\tau\Big]d\tau ds.$$

Since for $t \geq 0$, we have

$$u'(t) = AS(t)\varphi(0) + C(t)\varphi'(0) + \int_0^t C(t-s)f(s, u_s, u'_s)ds,$$

then for $t \in [0, a]$, we have

$$|u'(t) - w(t)| \leq \int_0^t |S(t-s)| |f(s, u_s, u'_s) - f(s, u_s, w_s)|ds$$
$$+ \int_0^t \int_0^s |C(s-\tau)| |f_1(\tau, u_\tau, u'_\tau) - f_1(\tau, u_\tau, w_\tau)|d\tau ds \quad (4.1)$$
$$+ \int_0^t \int_0^s |C(s-\tau)| |f_2(\tau, u_\tau, u'_\tau)u'_\tau - f_2(\tau, u_\tau, w_\tau)u'_\tau|d\tau ds$$
$$+ \int_0^t \int_0^s |C(s-\tau)| |f_3(\tau, u_\tau, u'_\tau)v_\tau - f_3(\tau, u_\tau, w_\tau)v_\tau|d\tau ds.$$

Let $H = \{u'_s, w_s : s \in [0, a]\}$. Then H is a compact set, it follows that f, f_1, f_2 and f_3 are globally lipschitz on H. Let c_1 be such that for $t \in [0, a]$ and $x, y \in H$, we have

$$|f(t, x, x') - f(t, y, y')| \leq c_1 \|x - y\|$$
$$|f_1(t, x, x') - f_1(t, y, y')| \leq c_1 \|x - y\|$$
$$|f_2(t, x, x') - f_2(t, y, y')| \leq c_1 \|x - y\|$$
$$|f_3(t, x, x') - f_3(t, y, y')| \leq c_1 \|x - y\|.$$

Consequently, using equation (4.1) we can find a positive constant $k(a)$ such that

$$\|u'_\tau - w_\tau\| \leq k(a) \int_0^t \|u'_s - w_s\|ds \text{ for } s \in [0, a],$$

which implies that $u' = w$. Consequently, we deduce that the mild solution is twice continuously differentiable from $[-r, a]$ to X and the function $t \to f(t, u_t, u'_t)$ is continuously differentiable on $[0, a]$. We deduce that u is a strict solution of equation (1.1) on $[0, a]$. This holds for any $a < b_\varphi$. ∎

5. APPLICATION

For illustration, we propose to study the existence of solutions for the following model

$$\begin{cases} \dfrac{\partial^2 z(t,x)}{\partial t^2} = \dfrac{\partial^2 z(t,x)}{\partial x^2} + \displaystyle\int_{-r}^{0} g(t, z(t+\theta, x), z'(t+\theta, x)) d\theta \text{ for } t \geq 0 \text{ and } x \in [0, \pi] \\ z(t, 0) = z(t, \pi) = 0 \text{ for } t \geq 0 \\ z'(t, 0) = z'(t, \pi) = 0 \text{ for } t > 0 \\ z(\theta, x) = \varphi_0(\theta, x) \text{ for } \theta \in [-r, 0] \text{ and } x \in [0, \pi], \\ z'(\theta, x) = \varphi'_0(\theta, x) \text{ for } \theta \in [-r, 0] \text{ and } x \in [0, \pi] \end{cases} \tag{5.1}$$

where $g : \mathbb{R} \times \mathbb{R} \times \mathbb{R} \to \mathbb{R}$ is continuous and there exists a positive constant L such that for $x, y, x_1, y_1 \in \mathbb{R}$

$$|g(t, x, y) - g(t, x_1, y_1)| \leq L(|x - x_1| + |y - y_1|).$$

For example, we can take $g(t, x, y) = e^{-t^2}\left[\sin\left(\dfrac{x}{2}\right) + \sin\left(\dfrac{y}{2}\right)\right]$ for $(\theta, x, y) \in \mathbb{R}^- \times \mathbb{R} \times \mathbb{R}$. We can see that $|g(t, x_1, y_1) - g(t, x_2, y_2)| \leq \dfrac{1}{2}(|x_1 - x_2| + |y_1 - y_2|)$. The function $\varphi_0 : [-r, 0] \times [0, \pi] \to \mathbb{R}$ will be specified later. To rewrite equation (5.1) in the abstract form, we introduce the space $X = L^2([0, \pi]; \mathbb{R})$, functions vanishing at 0 and π, equipped with the L^2 norm that is to say for all $x \in X$,

$$|x|_{L^2} = \left(\int_0^\pi |x(s)|^2 ds\right)^{\frac{1}{2}}.$$

Let $e_n(x) = \sqrt{\dfrac{2}{\pi}} \sin(nx)$, $x \in [0, \pi]$, $n \in \mathbb{N}^*$, then $(e_n)_{n \in \mathbb{N}^*}$ is an orthonormal base for X. Let $A : X \to X$ be defined by

$$Ay = y''$$
$$D(A) = \{y \in X : C(t)y \text{ is a twice continuously differentiable function of } t\},$$

then
$$Ay = \sum_{n=1}^{+\infty} -n^2(y, e_n)e_n, \ y \in D(A).$$

We define the phase space
$$\mathcal{C} = BUC^1([-r, 0]; X)$$
where $BUC^1([-r, 0]; X)$ is the space of bounded uniformly continuous differentiable functions from $[-r, 0]$ into X with the norm $|\varphi| = \sup_{-r \leq \theta \leq 0} |\varphi(\theta)|$ and let $f : \mathbb{R} \times \mathcal{C} \times \mathcal{C} \to X$ be defined by
$$f(t, \varphi, \varphi')(x) = \int_{-r}^{0} g(t, \varphi(\theta)(x), \varphi'(\theta)(x)) d\theta \ \text{ for } x \in [0, \pi] \text{ and } t \geq 0,$$
where $\varphi \in \mathcal{C}$ is defined by
$$\varphi(\theta)(x) = \varphi_0(\theta, x) \text{ for } \theta \leq 0 \text{ and } x \in [0, \pi].$$

Let us pose $v(t) = z(t, x)$. Then equation (5.1) takes the following abstract form
$$\begin{cases} v''(t) = Av(t) + f(t, v_t, v'_t) \text{ for } t \geq 0 \\ v_0 = \varphi \\ v'_0 = \varphi'. \end{cases} \quad (5.2)$$

It is well known that A is the infinitesimal generator of strongly continuous cosine family $C(t)$, $t \in \mathbb{R}$ in X given by
$$C(t)y = \sum_{n=1}^{+\infty} \cos nt (y, e_n) e_n, \ y \in X,$$

which implies that ($\mathbf{H_0}$) is satisfied. Let $\varphi, \psi \in C^1([-r, 0]; X)$, we have

$$\begin{aligned}
\left(\int_0^\pi |f(s, \varphi, \varphi') - f(s, \psi, \psi')|^2 ds\right)^{\frac{1}{2}} &\leq \frac{1}{2}\left(\int_0^\pi \left(\int_{-r}^0 (|\varphi(\theta)(x) - \psi(\theta)(x)| \right.\right. \\
&\quad \left.\left. + |\varphi'(\theta)(x) - \psi'(\theta)(x)|) d\theta\right)^2 ds\right)^{\frac{1}{2}} \\
&\leq \frac{1}{2} r \pi^{\frac{1}{2}} \sup_{\substack{-r < \theta \leq 0 \\ 0 \leq x \leq \pi}} (|\varphi(\theta)(x) - \psi(\theta)(x)| \\
&\quad + |\varphi'(\theta)(x) - \psi'(\theta)(x)|) \\
&\leq \frac{1}{2} r \pi^{\frac{1}{2}} \|\varphi - \psi\|.
\end{aligned}$$

Consequently the function f satifies the condition of Corrolary 3.7. Then equation (5.2) has a unique mild solution which is defined for $t \geq 0$. For the regularity, we make the following assumptions.

($\mathbf{H_2}$) $g \in C^1(\mathbb{R} \times \mathbb{R} \times \mathbb{R}; \mathbb{R})$, such that $\dfrac{\partial g}{\partial t}, \dfrac{\partial g}{\partial x}$ and $\dfrac{\partial g}{\partial y}$ are locally lipschitz continuous.

($\mathbf{H_3}$)

$$\begin{cases} \varphi \in C^3([-r, 0] \times [0, \pi]) \text{ such that } \varphi(0), \varphi''(0) \in D(A), \varphi'(0), \varphi^{(3)}(0) \in E \\ \dfrac{\partial^2}{\partial \theta^2} \varphi(0, x) = \dfrac{\partial^2}{\partial x^2} \varphi(0, x) + \int_{-r}^0 g(0, \varphi(\theta, x)) d\theta \text{ for } x \in [0, \pi], \\ \dfrac{\partial^3}{\partial \theta^3} \varphi(0)(x) = \dfrac{\partial^2}{\partial x^2} \varphi'(0, x) \text{ for } x \in [0, \pi]. \end{cases}$$

Lemma 5.1. *Under the above assumptions, equation (5.1) has a unique strict solution v and the solution u defined by $u(t, x, x') = v(t)(x, x')$ for $t \geq 0$ and $x \in [0, \pi]$ becomes a solution equation (5.1).*

ACKNOWLEDGMENTS

The author wish to thank the referee for his (her) careful reading and valuable remarks which improve the presentation of the paper.

CONCLUSION

In this paper we study the existence and regularity of solutions for some nonlinear with finite using the cosine family theory. Some results of this study when the delay is infinite will be presented in next works.

REFERENCES

[1] Travis C.C. and Webb G.F., Cosine families and abstract nonlinear second order differential equations, *Transaction of Acta Mathematica Academiae Seientiarum Hungaricae Tomus* 32 (3-4), 75-96, (1978).

[2] Bobrowski A. and Chojnacki W. *Cosine Families and Semigroups Really Differ.*

[3] Bochenek J, An abstract nonlinear second order differential equation, *Annales Polonici Mathemativi*, 155-166, (1991).

[4] Pazy A., *Semigroups of Linear Operators and Applications to Partial Differential Equations,* Springer (1983).

In: Recent Studies in Differential Equations
Editor: Henry Forster
ISBN: 978-1-53618-389-4
© 2020 Nova Science Publishers, Inc.

Chapter 3

OSCILLATION RESULTS FOR NONLINEAR NEUTRAL IMPULSIVE DIFFERENTIAL EQUATIONS

Shyam Sundar Santra*
Department of Mathematics, JIS College of Engineering, Kalyani, India

Abstract

This chapter presents sufficient conditions for oscillation of solutions of neutral impulsive differential equations (IDEs)

$$\begin{cases} \frac{d}{dt}\big[x(t) - r(t)x(t-\tau_1)\big] + q(t)H\big(x(t-\sigma_1)\big) = 0, & t \geq t_0,\ t \neq \tau_k \\ x(\tau_k^+) = J_k\big(x(\tau_k)\big), & k \in \mathbb{N} \\ x\big((\tau_k - \tau_1)^+\big) = J_k\big(x(\tau_k - \tau_1)\big), & k \in \mathbb{N} \end{cases}$$

for $|r(t)| < +\infty$.

Keywords: oscillation, nonoscillation, delay, impulsive differential equations, neutral

Mathematics Subject Classification 2010: 34K

---

*Corresponding Author's Email: shyam01.math@gmail.com.

1. INTRODUCTION

Gradual devolvement of natural phenomena in various realistic problems are observed, described and interpreted by IDEs. In this orientation we refer some monographs (see for e.g., [1,2]) on impulsive differential equations to the readers. Differential equations represent several processes which are analyzed in pure and applied sciences. Though, the situation is not so smooth in all physical occurrences which have a unpredicted change in their states like mechanical systems, biological systems such as blood flows, heart beats, chemical technology, biotechnology processes, theoretical physics, radio-physics, industrial robotics, mathematical economy, ecology, electric technology, metallurgy, population dynamics, chemistry, pharmacokinetics, engineering, medicine, control theory and continuous.

In [3], Greaf et al. have taken

$$\begin{cases} \frac{d}{dt}[x(t) - r(t)x(t - \tau_1)] + q(t)|x(t - \sigma_1)|^\lambda \mathrm{sgn}(x(t-\sigma_1)) = 0, \\ x(\tau_k^+) = b_k x(\tau_k), \quad t \geq t_0, t \neq \tau_k, k \in \mathbb{N}, \end{cases} \quad (1.1)$$

where $\tau_1, \lambda \in (0, \infty)$, $\sigma_1 \in [0, \infty)$, and developed sufficient conditions for oscillation of (1.1) when $r, q \in PC([t_0, \infty), \mathbb{R}^+)$. In [4], Shen and Zou have established sufficient conditions for oscillation of solutions of neutral IDEs of the form

$$\begin{cases} \frac{d}{dt}[x(t) - r(t)x(t - \tau_1)] + q_1(t)x(t - \sigma_2) - q_2(t)x(t - \sigma_3) = 0, t \geq t_0, \\ x(\tau_k^+) = J_k(x(\tau_k)), \quad t \neq \tau_k \quad k = 1, 2, \cdots \end{cases}$$
$$(1.2)$$

where $\tau_1 \in (0, \infty)$, $\sigma_2 \geq \sigma_3 > 0$, $r, q_1, q_2 \in PC([t_0, \infty), \mathbb{R}^+)$ and $b_k \leq \frac{J_k(y)}{y} \leq 1$ where $b_k's$ are positive real numbers. In [5], Karpuz et al. have generalised the work of [4] by taking nonhomogeneous systems of (1.2) only for $r \in PC([t_0, \infty), \mathbb{R}^+)$ and $b_k \leq \frac{J_k(y)}{y} \leq 1$. In [6], Santra and Tripathy have established sufficient conditions for oscillation of solution of

$$\begin{cases} \frac{d}{dt}[x(t) - r(t)x(t - \tau_1)] + q(t)H(x(t - \sigma_1)) = 0, \quad t \geq t_0, t \neq \tau_k \\ x(\tau_k^+) = J_k(x(\tau_k)), \quad k \in \mathbb{N} \\ x((\tau_k - \tau_1)^+) = J_k(x(\tau_k - \tau_1)), \quad k \in \mathbb{N} \end{cases}$$
$$(1.3)$$

for negative range of r and $1 \leq J_k(x)/x \leq b_k$, $b_k \leq J_k(x)/x \leq 1$ and $J_k(x)/x \leq b_k \leq 1$ where $b'_k s$ are positive real numbers, till now the problem is open for $a_k \leq J_k(x)/x \leq b_k$ where $a'_k s$ and $b'_k s$ are positive real numbers. More on neutral IDEs, we refer [7–14] and the references cited there in.

Motivated by the above works, an attempt is made hare to discuss the oscillatory properties of solutions of the IDEs (1.3) for the different ranges of the neutral coefficient r. The following conditions are required for further analysis.
*

(A1) $q \in C([t_0, \infty), \mathbb{R}^+)$, $\tau_1, \sigma_1 > 0$, $\{\tau_k\} \subset [t_0, \infty)$ is increasing and divergent;

(A2) $H \in C(\mathbb{R}, \mathbb{R})$ is nondecreasing with $vH(v) > 0$ for $v \neq 0$, and there exists $K > 0$ such that $|H(v)| \geq K|v|$, $v \in \mathbb{R}$, i.e.,

$$H(v) = v(K + |v|^\mu), \quad v \in \mathbb{R}, \ \mu > 0;$$

(A3) $r \in PC([t_0, \infty), \mathbb{R})$ and $J_k \in C(\mathbb{R}, \mathbb{R})$ is continuous, there exists numbers $a_k > 0$ and $b_k > 0$ such that $a_k \leq \frac{J_k(u)}{u} \leq b_k$ for $u \neq 0$ and $k \in \mathbb{N}$.

With (1.3), we apply an initial condition

$$x(t) = \varphi(t), \quad t \in [t_0 - \rho, t_0],$$

where $\rho := \max\{\tau_1, \sigma_1\}$, $\varphi \in PC([t_0 - \rho, t_0], \mathbb{R})$ is continuous on $[t_0, \infty) \cap (\tau_k, \tau_{k+1})$ such that $\varphi(\tau_k^+) := \lim_{t \to \tau_k^+} \varphi(t)$ exists and $\varphi(\tau_k) = \varphi(\tau_k^-) := \lim_{t \to \tau_k^-} \varphi(t)$.

A function x, real-valued, is termed a solution corresponding to t_0 of the initial value problem (1.3) if: *

(i) $x(t) = \phi(t)$ for $t_0 - \rho \leq t \leq t_0$, $x(t)$ is continuous for $t \geq t_0$ and $t \neq \tau_k$, $k = 1, 2, \cdots$;

(ii) $[x - p \times x(\cdot - \tau_1)]$ is continuously differentiable on $[t_0, \infty)$, $t \neq \tau_k$, $t \neq \tau_k + \tau_1$, $t \neq \tau_k + \sigma_1$, $k = 1, 2, \cdots$, and x satisfies the first equation in (1.3);

(iii) $x(\tau_k^+)$, $x(\tau_k^-)$, $x\big((\tau_k - \tau_1)^+\big)$ and $x\big((\tau_k - \tau_1)^-\big)$ exist, $x(\tau_k^-) = x(\tau_k)$, $x\big((\tau_k - \tau_1)^-\big) = x(\tau_k - \tau_1)$, and the last two equations in (1.3) are satisfied.

If a solution x of (1.3) is either eventually positive or negative then it is called nonoscillatory; otherwise, it is oscillatory.

2. PRELIMINARY RESULTS

Lemma 2.1. *[6] Let $r(t) \geq 0$ for $t \geq t_0$. Assuming (A1)–(A3) hold, a sequence $\{\zeta_k\}$ exists such that $\zeta_k \in (\tau_k, \tau_{k+1}]$, $\zeta_{k+1} - \zeta_k = \tau_1$, $r(\zeta_k) > 0$ for $k = 1, 2, \cdots$ and*

$$\sum_{i=1}^{\infty} \prod_{j=1}^{i} \frac{b_j}{r(\zeta_j)} = \infty.$$

Further assume that

$$\begin{cases} r(\tau_k^+) \geq b_k r(\tau_k), & \tau_k - \tau_1 \neq \tau_i, \ i < k \\ a_k r(\tau_k^+) \geq b_k r(\tau_k), & \tau_k - \tau_1 = \tau_i, \ i < k. \end{cases} \quad (2.1)$$

Let x be the solution of (1.3) such that $x(t) > 0$, $x(t - \rho) > 0$ for $t \geq t_0 + \rho$. Then, $y(\tau_k^+) \geq 0$ for $t \in (_k, \tau_{k+1}]$, $k \geq 0$. In addition, $y(t) > 0$ for $t \geq t_0$, where

$$y(t) = x(t) - r(t)x(t - \tau_1). \quad (2.2)$$

Proof. From (1.3), it follows that

$$y'(t) = -Kq(t)x(t - \sigma_1) \leq 0, \quad t \geq t_0, \ t \neq \tau_k \quad (2.3)$$

that is, y is nonincreasing on $[t_0, \infty)$. Because of (2.1) and (A3), and if $\tau_k - \tau_1 = \tau_i$ with $i < k$, then

$$\begin{aligned} y(\tau_k^+) &= x(\tau_k^+) - r(\tau_k^+)x\big((\tau_k - \tau_1)^+\big) \\ &= J_k\big(x(\tau_k)\big) - r(\tau_k^+)J_k\big(x(\tau_k - \tau_1)\big) \\ &\leq b_k x(\tau_k) - r(\tau_k^+)a_k x(\tau_k - \tau_1) \\ &\leq b_k x(\tau_k) - b_k r(\tau_k)x(\tau_k - \tau_1) = b_k y(\tau_k) \end{aligned}$$

and if $\tau_k - \tau_1 \neq \tau_i$ with $i < k$, then

$$\begin{aligned} y(\tau_k^+) &= x(\tau_k^+) - r(\tau_k^+)x\big((\tau_k - \tau_1)^+\big) \\ &= J_k\big(x(\tau_k)\big) - r(\tau_k^+)x(\tau_k - \tau_1) \\ &\leq b_k x(\tau_k) - b_k r(\tau_k)x(\tau_k - \tau_1) = b_k y(\tau_k). \end{aligned}$$

Therefore,
$$y(\tau_k^+) \le b_k y(\tau_k), \quad k = 1, 2, \cdots. \tag{2.4}$$

We assert that $y(\tau_k) \ge 0$ for $k = 1, 2, \cdots$. If not, there exists $m \ge 1$ such that $y(\tau_m) < 0$. As y is monotonically decreasing on $[t_0, \infty)$, then there exists $\mu > 0$ such that $y(\tau_m) = -\mu$. Consequently, $y(\tau_m^+) \le -b_m \mu$. Indeed, $y(t) \le y(\tau_m^+) \le -b_m \mu$ for $\tau_m < t \le \tau_{m+1}$. Further,

$$y(\tau_{m+1}^+) \le b_{m+1} y(\tau_{m+1}) \le b_{m+1} y(\tau_m^+) \le -b_m b_{m+1} \mu,$$

which implies

$$y(t) \le y(\tau_{m+1}^+) \le -b_m b_{m+1} \mu \quad \text{for } \tau_{m+1} < t \le \tau_{m+2}.$$

Going forward inductively, we get

$$y(t) \le y(\tau_{m+n+1}^+) \le -b_m b_{m+1} \cdots b_{m+n+1} \mu$$

for $\tau_{m+n} < t \le \tau_{m+n+1}$, $n = 0, 1, \cdots$. Now,

$$\begin{aligned}x(\zeta_{m+n}) &= y(\zeta_{m+n}) + r(\zeta_{m+n}) x(\zeta_{m+n-1}) \\ &\le y(\tau_{m+n}^+) + r(\zeta_{m+n}) x(\zeta_{m+n-1}) \\ &\le -b_m b_{m+1} \cdots b_{m+n} \mu + r(\zeta_{m+n}) x(\zeta_{m+n-1}).\end{aligned}$$

The aforementioned relation is applied recursively and we get

$$\begin{aligned}x(\zeta_{m+n}) \le &-\mu \big[b_m b_{m+1} \cdots b_{m+n} \\ &+ r(\zeta_{m+n}) b_m b_{m+1} \cdots b_{m+n-1} \\ &+ r(\zeta_{m+n}) r(\zeta_{m+n-1}) b_m b_{m+1} \cdots b_{m+n-2} \\ &+ \cdots \\ &+ r(\zeta_{m+n}) r(\zeta_{m+n-1}) \cdots r(\zeta_{m+2}) b_m b_{m+1} \big] \\ &+ r(\zeta_{m+n}) r(\zeta_{m+n-1}) \cdots r(\zeta_{m+1}) x(\zeta_m) \\ = &\prod_{j=1}^{m} r(\zeta_{m+j}) \left[x(\zeta_m) - \mu b_m \sum_{i=1}^{n} \prod_{j=1}^{i} \frac{b_{m+j}}{r(\zeta_{m+j})} \right].\end{aligned}$$

As a consequence, $x(\zeta_{m+n}) < 0$ for moderately large n, which is a contradiction. Hence, $y(\tau_k) \ge 0$ for $k = 1, 2, \cdots$. As $y(t) \ge y(\tau_1)$ for $t_0 \le t \le \tau_1$, we have $y(t_0) \geq 0$. In other hand, $y(t) \geq y(\tau_{k+1}) \geq 0$ for $\tau_k < t \leq \tau_{k+1}$ indicates that $y(\tau_k^+) \geq 0$ and as a result $y(t) \geq 0$ for $t \in (\tau_k, \tau_{k+1}]$, $k \geq 0$.

Finally, we have to show that $y(t) > 0$ for $t \geq t_0$. We claim that $y(\tau_k) > 0$, $k = 0, 1, \cdots$. If not, let there exists $m \geq 0$ such that $y(\tau_m) = 0$. After integrating (2.3) from τ_m to τ_{m+1}, we get

$$y(\tau_{m+1}) = y(\tau_m^+) - K \int_{\tau_m}^{\tau_{m+1}} q(\eta) x(\eta - \sigma_1) d\eta$$

$$\leq b_m y(\tau_m) - K \int_{\tau_m}^{\tau_{m+1}} q(\eta) x(\eta - \sigma_1) d\eta$$

$$= -K \int_{\tau_m}^{\tau_{m+1}} q(\eta) x(\eta - \sigma_1) d\eta,$$

contradicts. Eventually, $y(\tau_k) > 0$ for all large k. Ultimately, $y(t) > 0$ for all large $t \geq t_0$ as $y(t) \geq y(\tau_{k+1}) > 0$ for $t \in (\tau_k, \tau_{k+1}]$. This completes the proof of the lemma 2.1. □

3. Sufficient Conditions for Oscillation

In this section, sufficient conditions are obtained for the oscillatory behaviour of solution of (1.3).

Theorem 3.1. *Let $r(t) \geq 0$ for $t \geq t_0$. Assume that the conditions of Lemma 2.1 hold. Furthermore, assume that *

(A4) $\tau_{k+1} - \tau_k \geq T$, $k = 1, 2, \cdots$

*hold. If either *

(A5) $\limsup_{k \to \infty} \left(\frac{1}{b_k} \int_{\tau_k}^{\tau_k + T} q(\eta) d\eta \right) > \frac{1}{K}$ *if $\sigma_1 \geq T > 0$*

*or *

(A6) $\limsup_{k \to \infty} \left(\frac{1}{b_k} \int_{\tau_k}^{\tau_k + \sigma_1} q(\eta) d\eta \right) > \frac{1}{K}$ *if $0 < \sigma_1 < T$*

holds. Then, every solution of (1.3) is oscillatory.

Proof. Suppose on the contrary that x is a nonoscillatory solution of (1.3). Without loss of generality, we assume that $x(t) > 0$, $x(t - \rho) > 0$ for $t \geq t_0$. By Lemma 2.1, $y(t) \geq 0$ for $t \in (\tau_k, \tau_{k+1}]$, $k = 0, 1, \cdots$. We consider the following two possible cases.

Case 1. Let $\sigma_1 \geq T > 0$. Using $y(t) \leq x(t)$ in (2.3) and then integrating it from τ_k to $\tau_k + T$, we obtain

$$y(\tau_k + T) - y(\tau_k^+) + K \int_{\tau_k^+}^{\tau_k+T} q(\eta)y(\eta - \sigma_1)\mathrm{d}\eta \leq 0.$$

Indeed, $\eta - \sigma_1 \leq \tau_k + T - \sigma_1$ and $\sigma_1 \geq T$ implies that $\tau_k + T - \sigma_1 \leq \tau_k$ and hence the last inequality reduces to

$$y(\tau_k + T) - y(\tau_k^+) + Ky(\tau_k) \int_{\tau_k^+}^{\tau_k+T} q(\eta)\mathrm{d}\eta \leq 0.$$

Consequently, by (2.4), we get

$$y(\tau_k + T) - y(\tau_k^+) + K\frac{y(\tau_k^+)}{b_k} \int_{\tau_k^+}^{\tau_k+T} q(\eta)\mathrm{d}\eta \leq 0.$$

Therefore,

$$y(\tau_k + T) + y(\tau_k^+)\left[\frac{K}{b_k} \int_{\tau_k^+}^{\tau_k+T} q(\eta)\mathrm{d}\eta - 1\right] \leq 0$$

is not possible due to (A5).

Case 2. Let $0 < \sigma_1 < T$. Integrating (2.3) from τ_k to $\tau_k + \sigma_1$, we obtain

$$y(\tau_k + \sigma_1) - y(\tau_k^+) + K \int_{\tau_k^+}^{\tau_k+\sigma_1} q(\eta)y(\eta - \sigma_1)\mathrm{d}\eta \leq 0. \quad (3.1)$$

Using $\tau_k + \sigma_1 \geq \eta$, that is, $\eta - \sigma_1 \leq \tau_k$, it follows from (3.1) that

$$y(\tau_k + \sigma_1) - y(\tau_k^+) + Ky(\tau_k) \int_{\tau_k^+}^{\tau_k+\sigma_1} q(\eta)\mathrm{d}\eta \leq 0,$$

that is,

$$y(\tau_k + \sigma_1) + y(\tau_k^+)\left[\frac{K}{b_k} \int_{\tau_k^+}^{\tau_k+\sigma_1} q(\eta)\mathrm{d}\eta - 1\right] \leq 0,$$

which is not possible due to (A6).

This completes the proof of the theorem. $\square$

Theorem 3.2. *Let $-p_0 \leq r(t) \leq 0$ for $t \geq t_0$, where p_0 is a positive constant. Assume that (A1)–(A3) and (A4) hold. Furthermore, assume that* *

(A7) $Q(t) := \min\{q(t), q(t - \tau_1)\}, t \geq \tau_1;$

(A8) $\begin{cases} r\big((\tau_k - \tau_1)^+\big) \geq b_k r(\tau_k - \tau_1), & \tau_k - 2\tau_1 \neq \tau_i, \ i < k \\ r\big((\tau_k - \tau_1)^+\big) \geq r(\tau_k - \tau_1), & \tau_k - 2\tau_1 = \tau_i, \ i < k \end{cases}$

hold. If either *

(A9) $\limsup_{k \to \infty} \left(\frac{1}{b_k} \int_{\tau_k}^{\tau_k + T} Q(\eta) d\eta \right) > \frac{1+a}{K}$ *if $\sigma_1 \geq T > 0$ and $\sigma_1 - \tau_1 \geq T$*

or *

(A10) $\limsup_{k \to \infty} \left(\frac{1}{b_k} \int_{\tau_k}^{\tau_k + \tau_1} Q(\eta) d\eta \right) > \frac{1+a}{K}$ *if $2\tau_1 < \sigma_1 < T$*

holds. Then, every solution of (1.3) is oscillatory.

Proof. Proceeding as in the proof of Theorem 3.1, we get (2.3). Hence, y is nonincreasing on $[t_0, \infty)$. It is easy to see that

$$\begin{aligned} 0 &= y'(t) + Kq(t)x(t - \sigma_1) + ay'(t - \tau_1) + aKq(t - \tau_1)x(t - \tau_1 - \sigma_1) \\ &\geq y'(t) + ay'(t - \tau_1) + KQ(t)\big[x(t - \sigma_1) + ax(t - \tau_1 - \sigma_1)\big] \\ &\geq y'(t) + ay'(t - \tau_1) + KQ(t)y(t - \sigma_1). \end{aligned} \quad (3.2)$$

Case 1. Let $\sigma_1 \geq T > 0$ and $\sigma_1 - \tau_1 \geq T > 0$. Integrating (3.2) on $(\tau_k, \tau_k + T)$, we obtain

$$y(\tau_k + T) - y(\tau_k^+) + ay(\tau_k + T - \tau_1) - ay\big((\tau_k - \tau_1)^+\big) \\ + K \int_{\tau_k^+}^{\tau_k + T} Q(\eta) y(\eta - \sigma_1) d\eta \leq 0. \quad (3.3)$$

Using (2.1) and (2.3), it follows that y is eventually positive and nonincreasing. Indeed, $y(\tau_k + T) \leq y(\tau_k + T - \tau_1)$, $y(\tau_k^+) \leq y\big((\tau_k - \tau_1)^+\big)$ and $\eta \leq \tau_k + T$, imply that $\eta - \sigma_1 \leq \tau_k + T - \sigma_1 \leq \tau_k - \tau_1$ for which $y(\eta - \sigma_1) \geq y(\tau_k - \tau_1)$. Therefore, (3.3) becomes

$$(1 + a)y(\tau_k + T) - (1 + a)y\big((\tau_k - \tau_1)^+\big) + Ky(\tau_k - \tau_1) \int_{\tau_k^+}^{\tau_k + T} Q(\eta) d\eta \leq 0. \quad (3.4)$$

Furthermore, $\tau_k - 2\tau_1 \neq \tau_i$ for $i < k$ implies that

$$\begin{aligned}
y\big((\tau_k - \tau_1)^+\big) &= x\big((\tau_k - \tau_1)^+\big) - r\big((\tau_k - \tau_1)^+\big)x(\tau_k^+ - 2\tau_1) \\
&= J_k\big(x(\tau_k - \tau_1)\big) - r\big((\tau_k - \tau_1)^+\big)x(\tau_k - 2\tau_1) \\
&\leq b_k x(\tau_k - \tau_1) - b_k r(\tau_k - \tau_1)x(\tau_k - 2\tau_1) = b_k y(\tau_k - \tau_1)
\end{aligned}$$

due to (A3) and (A8), and $\tau_k - 2\tau_1 = \tau_i$ for $i < k$ implies that

$$\begin{aligned}
y\big((\tau_k - \tau_1)^+\big) &= x\big((\tau_k - \tau_1)^+\big) - r\big((\tau_k - \tau_1)^+\big)x(\tau_k^+ - 2\tau_1) \\
&= J_k\big(x(\tau_k - \tau_1)\big) - r\big((\tau_k - \tau_1)^+\big)J_k\big(x(\tau_k - 2\tau_1)\big) \\
&\leq b_k x(\tau_k - \tau_1) - r\big((\tau_k - \tau_1)^+\big)b_k x(\tau_k - 2\tau_1) \\
&\leq b_k x(\tau_k - \tau_1) - r(\tau_k - \tau_1)b_k x(\tau_k - 2\tau_1) = b_k y(\tau_k - \tau_1)
\end{aligned}$$

due to (A3) and (A8). Hence for all k,

$$y\big((\tau_k - \tau_1)^+\big) \leq b_k y(\tau_k - \tau_1). \tag{3.5}$$

Using (3.5) in (3.4), we get

$$(1+a)y(\tau_k + T) + y\big((\tau_k - \tau_1)^+\big)\left[\frac{K}{b_k}\int_{\tau_k^+}^{\tau_k + T} Q(\eta)\mathrm{d}\eta - (1+a)\right] \leq 0,$$

which is not possible due to (A9).

Case 2. Let $2\tau_1 < \sigma_1 < T$. Integrating (3.2) on $(\tau_k, \tau_k + \tau_1)$, we obtain

$$y(\tau_k + \tau_1) - y(\tau_k^+) + ay(\tau_k) - ay\big((\tau_k - \tau_1)^+\big) \\ + K\int_{\tau_k^+}^{\tau_k + \tau_1} Q(\eta)y(\eta - \sigma_1)\mathrm{d}\eta \leq 0,$$

that is,

$$y(\tau_k + \tau_1) - y\big((\tau_k - \tau_1)^+\big) + ay(\tau_k + \tau_1) - ay\big((\tau_k - \tau_1)^+\big) \\ + K\int_{\tau_k^+}^{\tau_k + \tau_1} Q(\eta)y(\eta - \sigma_1)\mathrm{d}\eta \leq 0.$$

Consequently,

$$(1+a)y(\tau_k + \tau_1) - (1+a)y\big((\tau_k - \tau_1)^+\big) \\ + K\int_{\tau_k^+}^{\tau_k + \tau_1} Q(\eta)y(\eta - \sigma_1)\mathrm{d}\eta \leq 0. \tag{3.6}$$

Indeed, $\eta \leq \tau_k + \tau_1$ implies that $\eta - \sigma_1 \leq \tau_k + \tau_1 - \sigma_1 \leq \tau_k - \tau_1$ and hence $y(\eta - \sigma_1) \geq y(\tau_k - \tau_1)$. Using (3.5) in (3.6) it follows that

$$(1+a)y(\tau_k + \tau_1) + y\Big((\tau_k - \tau_1)^+\Big)\Big[\frac{K}{b_k}\int_{\tau_k^+}^{\tau_k+\tau_1} Q(\eta)\mathrm{d}\eta - (1+a)\Big] \leq 0,$$

which is not possible due to (A10).

This completes the proof of the theorem. $\square$

Remark 3.1. *Theorem 3.1 and Theorem 3.2 hold true for* $a_k = 1$ *for all* k.

Remark 3.2. *Theorem 3.1 and Theorem 3.2 hold true for* $b_k = 1$ *for all* k.

REFERENCES

[1] Bainov, D. D. and Simeonov, P. S. (1995). Impulsive Differential Eqations: Asympotic Properties of the Solutions, 28, *Series on Advances in Mathematoics for Applied Sciences*, World Scientific, Singapure.

[2] Lakshmikantham, V., Bainov, D. D. and Simieonov, P. S. (1989). *Oscillation Theory of Impulsive Differential Equations,* World Scientific, Singapore.

[3] Greaf, J. R., Shen, J. H. and Stavroulakis, I. P. (2002). Oscillation of impulsive neutral delay differential equations, *J. Math. Anal. Appl.*, 268, 310–333.

[4] Shen, J. and Zou, Z. (2008). Oscillation criteria for first order impulsive differential equations with positive and negative coefficients, *J. Comp. Appl. Math.*, 217, 88–37.

[5] Karpuz, B. and Öcalan, Ö. (2012). Oscillation criteria for a first-order forced differential equations under impulsive effects, *Adv. Dyn. Syst. Appl.*, 7(2): 205–218.

[6] Santra, S. S. and Tripathy, A. K. (2019). On oscillatory first order nonlinear neutral differential equations with nonlinear impulses, *J. Appl. Math. Comput.*, 59(1-2): 257–270.

[7] Agarwal, R. P. and F. Karakoç. (2010). A survey on oscillation of impulsive delay differential equations, *Comput. Math. Appl.*, 60, 1648–1685.

[8] Berezansky, L. and Braverman, E. (1996). Oscillation of a linear delay impulsive differential equations, *Comm. Appl. Nonlinear Anal.*, 3, 61–77.

[9] Gopalsamy, K. and Zhang, B.G. (1989) On delay Differential Equations with impulses, *J. Math. Anal. Appl.*, 139, 110–122.

[10] Luo, Z. and Jing, Z. (2008). Periodic boundary value problem for first-order impulsive functional differential equations, *Comput. Math. Appl.*, 55, 2094–2107.

[11] Pandian, S. and Purushothaman, G. (2012). Oscillation of impulsive neutral differential equation with several positive and negative coefficients, *J. Math. Comput. Sci.*, 2, 241–254.

[12] Shen, J. H. (1996). The existence of nonoscillatory solutions of delay differential eqations with impulses, *Appl. Math. Comput.*, 77, 153–165.

[13] Tripathy, A. K. and Santra, S. S. (2015). Necessary and sufficient conditions for oscillation of a class of first order impulsive differential equations, *Func. Differ. Equ.*, 22(3-4): 149-167.

[14] Tripathy, A. K. and Santra, S. S. (2016). Pulsatile constant and charecterization of first order neutral impulsive differential equations, *Com. Appl. Anal.*, 20, 65–76.

[15] Hale, J. K. (1977). *Theory of Functional Differential Equations*, Spinger-Verlag, New York.

[16] Tripathy, A. K. (2014). Oscillation criteria for a class of first order neutral impulsive differential-difference equations, *J. Appl. Anal. Comput.*, 4, 89–101.

[17] Tripathy, A. K., Santra, S. S. and Pinelas, S. (2016). Necessary and sufficient condition for asymptotic behaviour of solutions of a class of first order impulsive systems, *Adv. Dyn. Syst. Appl.*, 11(2): 135–145.

[18] Tripathy, A. K. and Santra, S. S. (2018). Necessary and suffcient conditions for oscillation of a class of second order impulsive systems, *Differ. Equ. Dyn. Syst.*, (published online on 09 May 2018).

Chapter 4

FIRST-ORDER FORCED FUNCTIONAL DIFFERENTIAL EQUATIONS

Shyam Sundar Santra[*]
Department of Mathematics, JIS College of Engineering,
Kalyani, India

Abstract

In this chapter, oscillatory behaviour of the solutions of a class of nonlinear first-order neutral differential equations with several delays of the form

$$\bigl(x(t) + p(t)x(t-\tau)\bigr)' + \sum_{i=1}^{m} q_i(t) H\bigl(x(t-\sigma_i)\bigr) = f(t)$$

are studied. This problem is considered in various ranges of the neutral coefficient p. The main tools are Knaster-Tarski fixed point theorem and Banach's fixed point theorem. Examples are provide to show feasibility and effectiveness of the main results.

Keywords: oscillation, nonoscillation, non-linear, delay, neutral differential equations, Knaster-Tarski fixed point theorem, Banach's fixed point theorem

Mathematics Subject Classification 2010: 34C10, 34C15, 34K40

---

[*]Corresponding Author's Email: shyam01.math@gmail.com.

1. INTRODUCTION

An increasing interest in oscillation of solutions to functional differential equations during the last few decades has been stimulated by applications arising in engineering and natural sciences. The challenges that the new classes of such equations provide in these application areas. Equations involving delay, and those involving advance and a combination of both arise in the models on lossless transmission lines in high speed computers which are used to interconnect switching circuits. The construction of these models using delays is complemented by the mathematical investigation of nonlinear equations. Moreover, the delay differential equations play an important role in modelling virtually every physical, technical, or biological process, from celestial motion, to bridge design, to interactions between neurons. There has been many investigations into the oscillation and nonoscillation of first order nonlinear neutral delay differential equations (see, e.g., [1], [2], [3], [4], [5], [6]). However, the study of oscillatory behaviour of solutions of (E_1) has received much less attention, which is due to mainly to the technical difficulties arising in its analysis.

In [1], Ahmed et al. have studied the oscillation properties of a linear differential equations of the form

$$(E_1) \qquad (r(t)(x(t) + p(t)x(t - \tau)))' + q(t)x(t - \sigma) = 0,$$

for the cases $p(t) \leq -1$, $-1 \leq p(t) < 0$ and $p(t) \equiv p \neq \pm 1$ and established sufficient conditions so that every solution of (E_1) is oscillatory. Their method has made the proof unnecessarily complicated and applicable to only homogeneous equations. In an another paper [7], Ahmed et al. considered the first order nonlinear neutral delay differential equations with variable coefficients of the form

$$(E_2) \qquad [r(t)(a(t)x(t) + p(t)x(t - \tau))]' + q(t)f(x(t - \sigma)) = 0,$$

and obtained some new sufficient conditions for the oscillation of all solutions of (E_2) by employing the Riccati transformation. In [4], Graef et al. considered (E_2) when $a(t) = 1 = r(t)$ and developed some sufficient conditions for the oscillation of all solutions of (E_2). In [3], Elabbasy et al. have studied first-order nonlinear neutral delay differential equation of the form

$$(E_3) \qquad (x(t) - q(t)x(t - \tau))]' + f(t, x(\tau(t))) = 0,$$

and established oscillation criteria for all solutions of (E_3) for $q(t) \neq 1$.

In [2], Das and Misra have made an attempt to study the oscillation properties of a nonlinear differential equations of type

$$(E_4) \qquad (x(t) - px(t-\tau))' + q(t)H(x(t-\sigma)) = f(t),$$

where $0 \leq p < 1$, $f(t) > 0$, and H satisfies the generalized sublinear condition

$$\int_0^{\pm k} \frac{dt}{H(t)} < \infty,$$

for every positive constant k, and established necessary and sufficient conditions so that every solution of (E_4) either oscillates or tends to zero. Their method has made the proof unnecessarily complicated and does not allow $f \equiv 0$ and H to be superlinear. Thus their result is applicable to only strictly nonhomogeneous equations. In [8], Parhi and Rath considered (E_4) for $p = \pm 1$ and established sufficient conditions under which every solution of (E_4) either oscillates or tends to zero or $\pm\infty$ as $t \to \infty$. In another paper [9], Parhi and Rath considered (E_4) for different ranges of $p(t)$ and established necessary and sufficient conditions under which every solution of (E_4) either oscillates or tends to zero or $\pm\infty$ as $t \to \infty$.

Hence in this chapter, the author has made an attempt to establish the sufficient condition of a class of first-order forced nonlinear neutral delay differential equation

$$(x(t) + p(t)x(t-\tau))' + \sum_{i=1}^{m} q_i(t) H\big(x(t-\sigma_i)\big) = f(t), \qquad (1.1)$$

where

$$\tau, \sigma_i \in \mathbb{R}_+ = (0, +\infty), \ i = 1, 2, \cdots, m,$$

$$p \in C([0, \infty), \mathbb{R}), \ q \in (\mathbb{R}_+, \mathbb{R}_+), \ f \in C(\mathbb{R}, \mathbb{R}),$$

and H satisfies

$$H \in C(\mathbb{R}, \mathbb{R}) \text{ with } uH(u) > 0 \text{ for } u \neq 0.$$

The objective of this chapter is to study (1.1) under various ranges of $p(t)$. This chapter contains two main section. In the first section, sufficient conditions for oscillation of solutions of (1.1) is obtained for different ranges of the neutral coefficient $p(t)$. In the second section, an attempt is made to establish the necessary and sufficient conditions so that every solution of (1.1) converges to zero. Clearly, equations (E_1) and (E_4) are particular cases of equations the (1.1). Therefore, it is interesting to study the more general equations (1.1). Unlike the work in [1], [2], [3], [4], [7] and [8] an attempt is made here to establish sufficient conditions under which every solution or every bounded solution of (1.1) oscillates. Of course, the impact of forcing term is considered. keeping in view of the influence of forcing function, this work is separated for forced and unforced equations.

By a solution of (1.1) we understand a function $x \in C([-\rho, \infty), \mathbb{R})$ such that $x(t) + p(t)x(t-\tau)$ is once continuously differentiable and (1.1) is satisfied for $t \geq 0$, where $\rho = \max\{\tau, \sigma_i\}$ for $i = 1, 2, \cdots, m$, and $\sup\{|x(t)| : t \geq t_0\} > 0$ for every $t_0 \geq 0$. A solution of (1.1) is said to be oscillatory if it has arbitrarily large zeros; Otherwise, it is called nonoscillatory.

Remark 1.1. *When the domain is not specified explicitly, all functional inequalities considered in this paper are assumed to hold eventually, i.e., they are satisfied for all t large enough.*

Remark 1.2. *Due to the assumptions and the form of the equation (1.1), we can deal only with eventually positive solutions of equation (1.1).*

2. SUFFICIENT CONDITIONS FOR OSCILLATION

In this section, sufficient conditions are obtained for oscillation of solutions of the equation (1.1). We need the following assumptions for this work in the sequel.

(A_1) there exists $\lambda > 0$ such that $H(u) + H(v) \geq \lambda H(u+v)$ for $u, v > 0$;

(A_2) $H(uv) = H(u)H(v)$ for $u, v \in \mathbb{R}$;

(A_3) $H(-u) = -H(u)$ for $u \in \mathbb{R}$;

(A_4) there exists $F \in C(\mathbb{R}, \mathbb{R})$ such that $F(t)$ changes sign with

$$-\infty < \liminf_{t \to \infty} F(t) < 0 < \limsup_{t \to \infty} F(t) < \infty \text{ and } F'(t) = f(t);$$

(A_5) $F^+(t) = \max\{F(t), 0\}$, $F^-(t) = \max\{-F(t), 0\}$

(A_6) there exists $F \in C(\mathbb{R}, \mathbb{R})$ such that $F(t)$ changes sign with

$$\liminf_{t \to \infty} F(t) = -\infty, \limsup_{t \to \infty} F(t) = +\infty \text{ and } F'(t) = f(t).$$

Remark 2.1. *Assumption* (A_2) *implies that* (A_3) *Indeed,* $H(1)H(1) = H(1)$ *and* $H(1) > 0$ *imply that* $H(1) = 1$. *Further,* $H(-1)H(-1) = H(1) = 1$ *implies that* $(H(1))^2 = 1$. *Since* $H(-1) < 0$, *we conclude that* $H(-1) = -1$. *Hence,*

$$H(-u) = H(-1)H(-u) = -H(u).$$

On the other hand, $H(uv) = H(u)H(v)$ *for* $u > 0$ *and* $v > 0$ *and* $H(-u) = -H(u)$ *imply that* $H(xy) = H(x)H(y)$ *for every* $x, y \in \mathbb{R}$.

Remark 2.2. *We may note that if* $x(t)$ *is a solution of* (1.1), *then* $y(t) = -x(t)$ *is also a solution of* (1.1) *provided that* H *satisfies* (A_2) *or* (A_3).

Theorem 2.1. *Let* $p(t) \geq 0$, $t \in \mathbb{R}_+$. *If* (A_2) *and* (A_6) *hold, then every solution of the equation* (1.1) *is oscillatory.*

Proof. Suppose for contrary that $x(t)$ is a nonoscillatory solution of equation (1.1). Then there exists $t_0 \geq \rho$ such that $x(t) > 0$ or $x(t) < 0$ for $t \geq t_0$. Assume that $x(t) > 0$ for $t \geq t_0$. Setting

$$z(t) = x(t) + p(t)x(t - \tau), \tag{2.1}$$

and

$$w(t) = z(t) - F(t), \tag{2.2}$$

it follows from (1.1) that

$$w'(t) = -\sum_{i=1}^{m} q_i(t) H\big(x(t - \sigma_i)\big) \leq 0 \tag{2.3}$$

for $t \geq t_1 > t_0$. Consequently, $w(t)$ is nonincreasing on $[t_2, \infty)$, $t_2 > t_1$. Hence we have $w(t) < 0$ or $w(t) > 0$ for $t \geq t_2 > t_1$. Since $z(t) > 0$, then $w(t) < 0$ for $t \geq t_2$ implies that $\liminf_{t \to \infty} F(t) \geq 0$ for $t \geq t_2$, a contradiction to (A_6). Hence, $w(t) > 0$ for $t \geq t_2$, then $\lim_{t \to \infty} w(t)$ exists. Writing

$$z(t) = w(t) + F(t),$$

we notice that
$$\begin{aligned}0 \le \liminf_{t\to\infty} z(t) &= \liminf_{t\to\infty}(w(t)+F(t))\\ &\le \limsup_{t\to\infty} w(t) + \liminf_{t\to\infty} F(t)\\ &= \lim_{t\to\infty} w(t) + \liminf_{t\to\infty} F(t) = -\infty,\end{aligned}$$

a contradiction due to (A_6).

If $x(t) < 0$ for $t \ge t_0$, then we set $y(t) = -x(t)$ for $t \ge t_0$ in (1.1) and we find

$$(y(t) + p(t)y(t-\tau))' + \sum_{i=1}^{m} q_i(t) H\big(y(t-\sigma_i)\big) = \widetilde{f}(t), \quad (2.4)$$

where $\widetilde{f}(t) = -f(t)$ due to (A_2). Let $\widetilde{F}(t) = -F(t)$. Then

$$-\infty < \liminf_{t\to\infty} \widetilde{F}(t) < 0 < \limsup_{t\to\infty} \widetilde{F}(t) < \infty$$

and $\widetilde{F}'(t) = \widetilde{f}(t)$ hold. Hence proceeding as above, we find a contradiction to (A_6). This completes the proof of the theorem. □

Theorem 2.2. *Let* $0 \le p(t) \le p < \infty$, $t \in \mathbb{R}_+$. *Assume that* (A_1), (A_2), (A_4) *and* (A_5) *hold. Furthermore, assume that*

(A_7) $\int_\rho^\infty \sum_{i=1}^m Q_i(\eta) H\big(F^+(\eta-\sigma_i)\big) dt = \infty = \int_\rho^\infty \sum_{i=1}^m Q_i(\eta) H\big(F^-(\eta-\sigma_i)\big) d\eta$

holds, then conclusion of the Theorem 2.1 is true, where for $t > \tau$, $Q_i(t) = \min\{q_i(t), q_i(t-\tau)\}$; $i = 1, 2, \cdots, m$.

Proof. On the contrary, we proceed as in the proof of the Theorem 2.1 to obtain that $w(t)$ is monotonic on $[t_2, \infty)$, $t_2 > t_1$. Since $z(t) > 0$, then $w(t) < 0$ for $t \ge t_2$ implies that $F(t) > 0$ for $t \ge t_2$, a contradiction to (A_4). Hence, $w(t) > 0$ for $t \ge t_2$. Ultimately, $z(t) > F(t)$ and hence $z(t) > \max\{0, F(t)\} = F^+(t)$ for $t \ge t_2$. Note that $\lim_{t\to\infty} w(t)$ exists. Due to (2.2), (1.1) becomes

$$0 = w'(t) + \sum_{i=1}^{m} q_i(t) H\big(x(t-\sigma_i)\big)$$
$$+ H(p)\Big[w'(t-\tau) + \sum_{i=1}^{m} q_i(t-\tau) H(x(t-\tau-\sigma_i))\Big]$$

for $t \geq t_2$ and because of (A_1) and (A_2), we find that

$$0 \geq w'(t) + H(p)w'(t-\tau) + \sum_{i=1}^{m} Q_i(t)\Big[H\big(x(t-\sigma_i)\big) + H\big(p\,x(t-\tau-\sigma_i)\big)\Big]$$

$$\geq w'(t) + H(p)w'(t-\tau) + \lambda \sum_{i=1}^{m} Q_i(t) H\big(z(t-\sigma_i)\big)$$

$$\geq w'(t) + H(p)w'(t-\tau) + \lambda \sum_{i=1}^{m} Q_i(t) H\big(F^+(t-\sigma_i)\big), \tag{2.5}$$

for $t \geq t_3 > t_2$. Integrating (2.5) from t_3 to $t(>t_3)$, we obtain

$$\lambda \int_{t_3}^{t} \sum_{i=1}^{m} Q_i(\eta) H\big(F^+(\eta-\sigma_i)\big) d\eta \leq -[w(\eta) + H(p)w(\eta-\tau)]_{t_3}^{\eta} < \infty, \text{ as } t \to \infty,$$

which is a contradiction to (A_7).

If $x(t) < 0$ for $t \geq t_0$, then we set $y(t) = -x(t)$ to obtain $y(t) > 0$ for $t \geq t_0$ and hence using equation (2.4), we obtain a contradiction due to (A_7). This completes the proof of the theorem. □

Theorem 2.3. *Let $-1 \leq p(t) \leq 0$, $t \in \mathbb{R}_+$. Suppose that (A_2), (A_4) and (A_5) hold. If any one of the following conditions*

(A_8) $\int_{\rho}^{\infty} \sum_{i=1}^{m} q_i(\eta) H\big(F^+(\eta-\sigma_i)\big) d\eta = \infty = \int_{\rho}^{\infty} \sum_{i=1}^{m} q_i(\eta) H\big(F^-(\eta+\tau-\sigma_i)\big) d\eta$

and

(A_9) $\int_{\rho}^{\infty} \sum_{i=1}^{m} q_i(\eta) H\big(F^-(\eta-\sigma_i)\big) d\eta = \infty = \int_{\rho}^{\infty} \sum_{i=1}^{m} q_i(\eta) H\big(F^+(\eta+\tau-\sigma_i)\big) d\eta$

holds, then conclusion of the Theorem 2.1 is true.

Proof. On the contrary, we proceed as in the proof of the Theorem 2.1 to obtain that $w(t)$ is monotonic on $[t_2, \infty)$, $t_2 > t_1$. If $w(t) < 0$ for $t \geq t_2$, then $z(t) < F(t)$ is a contradiction due to (A_4) when $z(t) > 0$. Ultimately, $z(t) < 0$ and $z(t) < F(t)$ for $t \geq t_3 > t_2$. Using the fact that $z(t) < 0$ for $t \geq t_3$, it follows that

$$x(t) \leq -p(t)x(t-\tau) \leq x(t-\tau) \leq x(t-2\tau) \leq \cdots \leq x(t_3),$$

implies that $x(t)$ is bounded on $[t_3, \infty)$. Consequently, $\lim_{t \to \infty} w(t)$ exists. Clearly, $-z(t) > -F(t)$ implies that $-z(t) > \max\{0, -F(t)\} = F^-(t)$. Therefore, for $t \geq t_3 > t_2$

$$-x(t - \tau) \leq p(t)x(t - \tau) \leq z(t) < -F^-(t)$$

gives rise to $x(t - \sigma_i) > F^-(t + \tau - \sigma_i)$, $t \geq t_4 > t_3$ for $i = 1, 2, \cdots, m$ and hence (2.3) reduced to

$$w'(t) + \sum_{i=1}^{m} q_i(t) H\left(F^-(t + \tau - \sigma_i)\right) \leq 0,$$

for $t \geq t_4 > t_3$. Integrating the last inequality from t_4 to ∞, we obtain

$$\int_{t_4}^{\infty} \sum_{i=1}^{m} q_i(\eta) H\left(F^-(\eta + \tau - \sigma_i)\right) d\eta < \infty,$$

which contradicts (A_8). Hence $w(t) > 0$ for $t \geq t_2 > t_1$. We note that $z(t) > F(t)$ and $z(t) < 0$ is not possible due to (A_4). Therefore $z(t) > 0$ and $z(t) \leq x(t)$ for $t \geq t_3 > t_2$. In this case, $\lim_{t \to \infty} w(t)$ exists. Because, it happens that $z(t) > F^+(t)$ for $t \geq t_3 > t_2$, then (2.3) can be viewed as

$$w'(t) + \sum_{i=1}^{m} q_i(t) H(F^+(t - \sigma_i)) \leq 0.$$

Integrating the last inequality from t_3 to ∞, we obtain

$$\int_{t_3}^{\infty} \sum_{i=1}^{m} q_i(s) H(F^+(s - \sigma_i)) ds < \infty,$$

a contradiction to (A_8). The case $x(t) < 0$ for $t \geq t_0$ is similar. Hence, the theorem is proved. □

Theorem 2.4. *Let $-\infty < -p \leq p(t) \leq -1$, $t \in \mathbb{R}_+$ and $p > 0$. If all conditions of Theorem 2.3 are satisfied, then every bounded solution of (1.1) oscillates.*

Proof. The proof of the theorem can be followed from the proof of the Theorem 2.3. Hence the details are omitted. □

Remark 2.3. In Theorem 2.2-2.4, H could be linear, sublinear or superlinear.

Theorem 2.5. Let $-\infty < -p \leq p(t) \leq -1$, $t \in \mathbb{R}_+$, $p > 0$ and $\tau \geq \sigma_i$ for $i = 1, 2, \cdots, m$. Assume that (A_2), (A_4), (A_5) and (A_8) hold. Furthermore, assume that

(A_{10}) $\dfrac{H(u)}{u^\beta} \geq \dfrac{H(v)}{v^\beta}$, $u \geq v > 0$, $\beta > 1$

and

(A_{11}) $\int_\rho^\infty \sum_{i=1}^m \dfrac{q_i(\eta) H\left(F^+(\eta+\tau-\sigma_i)\right)}{\left[F^+(\eta+\tau-\sigma_i)\right]^\beta} d\eta = \infty = \int_\rho^\infty \sum_{i=1}^m \dfrac{q_i(\eta) H\left(F^-(\eta+\tau-\sigma_i)\right)}{\left[F^-(\eta+\tau-\sigma_i)\right]^\beta} d\eta$

hold. Then conclusion of the Theorem 2.1 is true.

Proof. The proof of the theorem follows from the proof of Theorem 2.3 except the case when $w(t) < 0$, $z(t) < 0$ for $t \geq t_3 > t_2$. Since $z(t) \geq p(t)x(t-\tau)$, then

$$w(t) = z(t) - F(t) \geq p(t)x(t-\tau) - F(t), \quad t \geq t_3$$

implies that $w(t) - p(t)x(t-\tau) \geq -F(t)$ for $t \geq t_3$. Clearly, $w(t) - p(t)x(t-\tau) < 0$ is not possible due to (A_4) and the fact that $w(t) - p(t)x(t-\tau) = x(t) - F(t) \geq -F(t)$ if and only if $x(t) > 0$ for $t \geq t_3$. Ultimately, $w(t) - p(t)x(t-\tau) > 0$ and hence

$$w(t) - p(t)x(t-\tau) \geq \max\{0, -F(t)\} = F^-(t),$$

that is,

$$w(t) \geq p(t)x(t-\tau) + F^-(t) \geq -px(t-\tau) + F^-(t) > -px(t-\tau) \quad (2.6)$$

for $t \geq t_4 > t_3$. Since $w(t)$ is decreasing and $\tau \geq \sigma_i$ for $i = 1, 2, \cdots, m$, then it follows that

$$-w(t) \leq -w(t+\tau-\sigma_i) < px(t-\sigma_i), \quad t \geq t_4, \ i = 1, 2, \cdots, m.$$

Therefore,

$$\dfrac{H\left(x(t-\sigma_i)\right)}{[-w(t)]^\beta} \geq \dfrac{H\left(x(t-\sigma_i)\right)}{p^\beta x^\beta(t-\sigma_i)}, \quad t \geq t_4, i = 1, 2, \cdots, m. \quad (2.7)$$

Consequently,

$$-\frac{d}{dt}[-w(t)]^{1-\beta} = -(1-\beta)[-w(t)]^{-\beta}[-w'(t)]$$

$$= (\beta-1)[-w(t)]^{-\beta}\sum_{i=1}^{m}q_i(t)H(x(t-\sigma_i))$$

$$\geq (\beta-1)\sum_{i=1}^{m}q_i(t)\frac{H(x(t-\sigma_i))}{p^\beta x^\beta(t-\sigma_i)}, \quad t \geq t_4$$

due to (2.3) and (2.7). We may note from (2.6) that $0 > w(t) > -px(t-\tau) + F^-(t)$ implies that $x(t-\sigma_i) > p^{-1}F^-(t+\tau-\sigma_i)$ for $i = 1, 2, \cdots, m$ and hence

$$-\frac{d}{dt}[-w(t)]^{1-\beta} \geq (\beta-1)\sum_{i=1}^{m}q_i(t)\frac{H(p^{-1}F^-(t+\tau-\sigma_i))}{p^\beta[p^{-1}F^-(t+\tau-\sigma_i)]^\beta}, \quad (2.8)$$

for $t \geq t_4$ due to (A_{13}). Integrating (2.8) from t_4 to t, we get

$$(\beta-1)H(p^{-1})\int_{t_4}^{t}\sum_{i=1}^{m}q_i(\eta)\frac{H(F^-(\eta+\tau-\sigma_i))}{[F^-(\eta+\tau-\sigma_i)]^\beta}d\eta \leq -\big[-w(s)^{1-\beta}\big]_{t_4}^{t} < \infty, \text{ as } t \to \infty,$$

due to (A_2), which is a contradiction to (A_{11}).

The case $x(t) < 0$ for $t \geq t_0$ can similarly be dealt with. Hence the theorem is proved. □

3. NECESSARY AND SUFFICIENT CONDITIONS FOR OSCILLATION

In this section, we established the necessary and sufficient condition for asymptotic behavior of solutions of (1.1). We need the following assumptions for this section in the sequel:

(A_{12}) there exists $F \in C(\mathbb{R},\mathbb{R})$ such that $f(t) = F'(t)$ and F satisfy the condition $\lim_{t\to\infty}F(t) = M, |M| < \infty$

(A_{13}) $\int_0^\infty \sum_{i=1}^{m}q_i(\eta)d\eta = \infty$.

Lemma 3.1. *[10] Let $p, x, z \in C([0, \infty), \mathbb{R})$ be such that $z(t) = x(t) + p(t)x(t - \tau)$, $t \geq \tau > 0$, $x(t) > 0$ for $t \geq t_1 > \tau$, $\liminf_{t \to \infty} x(t) = 0$ and $\lim_{t \to \infty} z(t) = L \in \mathbb{R}$ exists. Let $p(t)$ satisfy one of the following conditions:*

$$i) \ 0 \leq p_1 \leq p(t) \leq p_2 < 1,$$
$$ii) \ 1 < p_3 \leq p(t) \leq p_4 < \infty,$$
$$iii) \ -\infty < -r_5 \leq p(t) \leq 0.$$

Then $L = 0$.

Remark 3.1. *If, in the above lemma, $x(t) < 0$ for $t \geq \tau > 0$, $\limsup_{t \to \infty} x(t) = 0$ and $\lim_{t \to \infty} z(t) = L \in \mathbb{R}$, exists, then $L = 0$.*

Theorem 3.1. *Assume that (A_1), (A_2), (A_3) hold and $0 \leq p_1 \leq p(t) \leq p_2 < 1$ for $t \in \mathbb{R}_+$. Let H be Lipschitzian on intervals of the form $[\alpha, \beta]$, $0 < \alpha < \beta < \infty$. Then every solution of (1.1) converges to zero as $t \to \infty$ if and only if (A_{13}) holds.*

Proof. Suppose that (A_{13}) holds. Let $x(t)$ be a solution of (1.1) on $[t_y, \infty)$, $t_y \geq 0$, then there exists $t_0 \geq \rho$ such that $x(t) > 0$ or $x(t) < 0$ for $t \geq t_0$. Assume that $x(t) > 0$ for $t \geq t_0$. Setting

$$z(t) = x(t) + p(t)x(t - \tau), \tag{3.1}$$

and

$$w(t) = z(t) - F(t), \tag{3.2}$$

it follows from (1.1) that

$$w'(t) = -\sum_{i=1}^{m} q_i(t) H(x(t - \sigma_i)) < 0 \tag{3.3}$$

for $t \geq t_1 > t_0$. Consequently, $w(t)$ is nonincreasing on $[t_2, \infty)$, $t_2 > t_1$. Hence we have $w(t) < 0$ or $w(t) > 0$ for $t \geq t_2 > t_1$.

Case 1. Let $w(t) > 0$ for $t \geq t_2$. So, $\lim_{t \to \infty} w(t)$ exists. We claim that $x(t)$ is bounded. If not, there exists $\{\zeta_n\}$ such that $\zeta_n \to \infty$ as $n \to \infty$, $x(\zeta_n) \to \infty$ as $n \to \infty$ and

$$x(\zeta_n) = \max\{x(\eta) : t_2 \leq \eta \leq \zeta_n\}.$$

Therefore,
$$w(\zeta_n) = x(\zeta_n) + p(\zeta_n)x(\zeta_n - \tau) - F(\zeta_n)$$
$$\geq (1 + p_1)x(\zeta_n - \tau) - F(\zeta_n),$$

yields $\lim_{t \to \infty} w(t) = \infty$, which is a contradiction. So, our claim holds.

Case 2. Let $w(t) < 0$ for $t \geq t_2$. In this case also $x(t)$ is bounded, because if $x(t)$ unbounded, then preceding as above we obtain $w(\eta_n) > 0$, which is a contradiction.

Hence, for any $w(t)$, $\lim_{t \to \infty} z(t)$ and $\lim_{t \to \infty} x(t)$ exists. Consequently, $w(t)$ is bounded and $\lim_{t \to \infty} w(t)$ exists. We claim that $\liminf_{t \to \infty} x(t) = 0$. If not, then there exists $t_3 > t_2$ and $\beta > 0$ such that $x(t - \sigma_i(t)) \geq \beta > 0$ for $t \geq t_3$. Ultimately,

$$\int_{t_3}^{t} \sum_{i=1}^{m} q_i(\eta) H(x(\eta - \sigma_i)) d\eta \geq H(\beta) \left[\int_{t_3}^{t} \sum_{i=1}^{m} q_i(\eta) d\eta \right] \to +\infty,$$

as $t \to \infty$, due to (A_{13}). On the other hand, integrating (3.3) from t_3 to $t(> t_3)$ we obtain

$$\int_{t_3}^{t} \sum_{i=1}^{m} q_i(\eta) H(x(\eta - \sigma_i)) d\eta = -[w(\eta)]_{t_3}^{t} < \infty, \text{ as } t \to \infty,$$

and this way we get a contradiction. So, our claim hold. Consequently, $\lim_{t \to \infty} z(t) = 0$ due to Lemma 3.1. As a result,

$$0 = \lim_{t \to \infty} z(t) = \limsup_{t \to \infty} (x(t) + p(t)x(t - \tau))$$
$$\geq \limsup_{t \to \infty} x(t)$$

and consequently $\limsup_{t \to \infty} x(t) = 0$, and hence $\liminf_{t \to \infty} x(t) = 0$. Consequently, $\lim_{t \to \infty} x(t) = 0$. An equivalent procedure can be followed for $x(t) < 0$ for $t \geq t_y$ to show that $\lim_{t \to \infty} x(t) = 0$.

In order to prove that the condition (A_{13}) is necessary, we assume that

$$\int_{0}^{\infty} \sum_{i=1}^{m} q_i(\eta) d\eta < \infty \tag{3.4}$$

and we need to show that the (1.1) admits a nonoscillatory solution which doesn't tend to zero as $t \to \infty$ when the limit exists. If possible, let there exist $t_1 > 0$ such that

$$\int_{t_1}^{\infty} \sum_{i=1}^{m} q_i(\eta) d\eta < \frac{1-p_2}{5K},$$

where $K = \max\{K_1, H(1)\}$, K_1 is the Lipschitz constant of H on $[\frac{1-p_2}{10}, 1]$. By (A_2), let $\lim_{t \to \infty} F(t) = M$. Then we can find $t_2 > t_1$ such that $|F(t) - M| < \frac{1-p_2}{10}$ for $t \geq t_2$. For $t_3 > \max\{t_1, t_2\}$, we set $Y = BC([t_3, \infty), \mathbb{R})$, the space of real valued bounded continuous functions on $[t_3, \infty)$. Clearly, Y is a Banach space with respect to supremum norm defined by

$$\|y\| = \sup\{|y(t)| : t \geq t_3\}.$$

Let's define

$$S = \left\{ u \in Y : \frac{1-p_2}{10} \leq u(t) \leq 1, \ t \geq t_3 \right\}.$$

Clearly, S is a closed and convex subspace of Y. Let $T : S \to S$ be defined by

$$Tx(t) = \begin{cases} Tx(t_3 + \rho), & t \in [t_3, t_3 + \rho] \\ -p(t)x(t-\tau) + \frac{1+4p_2}{5} + (F(t) - M) + \int_t^{\infty} \sum_{i=1}^{m} q_i(\eta) H(x(\eta - \sigma_i)) d\eta, & t \geq t_3 + \rho. \end{cases}$$

For every $x \in S$,

$$Tx(t) \leq \frac{1-p_2}{10} + \frac{1+4p_2}{5} + H(1)\left[\int_t^{\infty} \sum_{i=1}^{m} q_i(\eta) d\eta\right]$$

$$< \frac{1-p_2}{10} + \frac{1+4p_2}{5} + \frac{1-p_2}{5} = \frac{1+p_2}{2} < 1$$

and

$$Tx(t) \geq -p(t)x(t-\tau) + \frac{1+4p_2}{5} + (F(t) - M)$$

$$\geq -p_2 + \frac{1+4p_2}{5} - \frac{1-p_2}{10} = \frac{1-p_2}{10}$$

implies that $Tx \in S$. Now, for $y_1, y_2 \in S$

$$|Ty_1(t) - Ty_2(t)| \leq |p(t)||y_1(t-\tau) - y_2(t-\tau(t))|$$
$$+ K_1 \int_t^\infty \sum_{i=1}^m q_i(\eta)|y_1(\eta - \sigma_i) - y_2(\eta - \sigma_i)|d\eta$$
$$\leq p_2||y_1 - y_2|| + K_1||y_1 - y_2||\left[\int_t^\infty \sum_{i=1}^m q_i(\eta)d\eta\right]$$
$$< \left(p_2 + \frac{1-p_2}{5}\right)||y_1 - y_2||$$

implies that,

$$||Ty_1 - Ty_2|| \leq \mu||y_1 - y_2||,$$

that is, T is a contraction mapping, where $\mu = \frac{1+4p_2}{5} < 1$. Since S is complete and T is a contraction on S, then by the Banach's fixed point theorem T has a unique fixed point on $\left[\frac{1-p_2}{10}, 1\right]$. Hence $Tx = x$ and

$$x(t) = \begin{cases} x(t_3 + \rho), & t \in [t_3, t_3 + \rho] \\ -p(t)x(t-\tau) + \frac{1+4p_2}{5} + (F(t) - M) + \int_t^\infty \sum_{i=1}^m q_i(\eta) H(x(\eta - \sigma_i))d\eta, & t \geq t_3 + \rho \end{cases}$$

is a nonoscillatory solution of the system (1.1) on $\left[\frac{1-p_2}{10}, 1\right]$ such that $\lim_{t\to\infty} x(t) \neq 0$. Therefore, (A_{13}) is necessary. This completes the proof of the theorem. $\square$

Theorem 3.2. *Assume that* (A_1), (A_2), (A_3) *hold and* $1 < p_3 \leq p(t) \leq p_4 < \infty$ *such that* $p_3^2 > p_4$ *for* $t \in \mathbb{R}_+$. *Suppose that* H *is Lipschitzian on intervals of the form* $[\alpha, \beta]$, $0 < \alpha < \beta < \infty$. *Then every solution of* (1.1) *converges to zero as* $t \to \infty$ *if and only if* (A_{13}) *holds.*

Proof. Sufficient part is same as in the proof of Theorem 3.1. For the necessary part, we suppose that (3.4) holds. It is possible to find a $t_1 > 0$ such that

$$\int_{t_1}^\infty \sum_{i=1}^m q_i(\eta)d\eta < \frac{p_3 - 1}{2K},$$

where $K = \max\{K_1, K_2\}$, K_1 is the Lipschitz constant of H on $[a, b]$ and $K_2 = H(b)$ such that

$$a = \frac{2\mu(p_3^2 - p_4) - p_4(p_3 + p_3^2 - 2)}{2p_3^2 p_4}$$

$$b = \frac{p_3 - 1 + \mu}{p_3}, \quad \mu > \frac{p_4(p_3 + p_3^2 - 2)}{2(p_3^2 - p_4)} > 0.$$

Also, we can find $t_2 > 0$ such that $|F(t) - M| < \frac{1}{2}(p_3 - 1)$ for $t \geq t_2 > t_1$. Let $Y = BC([t_2, \infty), \mathbb{R})$ be the space of real valued bounded continuous functions on $[t_2, \infty)$. Clearly, Y is a Banach space with respect to supremum norm defined by

$$\|y\| = \sup\{|y(t)| : t \geq t_2\}.$$

Define

$$S = \left\{ u \in Y : a \leq u(t) \leq b, \ t \geq t_2 \right\}.$$

It is easy to verify that S is a closed convex subspace of Y. Let $T : S \to S$ be such that

$$Tx(t) = \begin{cases} Tx(t_2 + \rho), & t \in [t_2, t_2 + \rho] \\ -\frac{x(t+\tau)}{p(t+\tau)} + \frac{F(t+\tau) - M}{p(t+\tau)} + \frac{\mu}{p(t+\tau)} + \frac{1}{p(t+\tau)} \left[\int_{t+\tau}^{\infty} \sum_{i=1}^{m} q_i(\eta) H(x(\eta - \sigma_i)) d\eta \right], & t \geq t_2 + \rho. \end{cases}$$

For every $x \in S$,

$$Tx(t) \leq \frac{H(b)}{p(t+\tau)} \left[\int_{t+\tau}^{\infty} \sum_{i=1}^{m} q_i(\eta) d\eta \right] + \frac{p_3 - 1}{2p(t+\tau)} + \frac{\mu}{p(t+\tau)}$$

$$\leq \frac{1}{p_3} \left[\frac{2(p_3 - 1)}{2} + \mu \right] = b$$

and

$$Tx(t) \geq -\frac{x(t+\tau)}{p(t+\tau)} + \frac{F(t+\tau) - M}{p(t+\tau)} + \frac{\mu}{p(t+\tau)}$$

$$> -\frac{b}{p_3} - \frac{p_3 - 1}{2p_3} + \frac{\mu}{p_4}$$

$$= -\frac{p_3 - 1 + \mu}{p_3^2} - \frac{p_3 - 1}{2p_3} + \frac{\mu}{p_4}$$

$$= \frac{2\mu(p_3^2 - p_4) - p_4(p_3 - 2 + p_3^2)}{2p_3^2 p_4} = a$$

implies that $Tx \in S$. For $y_1, y_2 \in S$

$$|Ty_1(t) - Ty_2(t)| \leq \frac{1}{|p(t+\tau)|}|y_1(t+\tau) - y_2(t+\tau)|$$
$$+ \frac{K_1}{|p(t+\tau)|}\left[\int_{t+\tau}^{\infty}\sum_{i=1}^{m}q_i(\eta)|y_1(\eta-\sigma_i) - y_2(\eta-\sigma_i)|d\eta\right]$$
$$\leq \frac{1}{p_3}||y_1 - y_2|| + \frac{K_1}{p_3}||y_1 - y_2||\left[\int_{t+\tau}^{\infty}\sum_{i=1}^{m}q_i(\eta)d\eta\right]$$
$$< \left(\frac{1}{p_3} + \frac{p_3-1}{2p_3}\right)||y_1 - y_2||,$$

implies that,
$$||Ty_1 - Ty_2|| \leq \mu ||y_1 - y_2||,$$

that is, T is a contraction, where $\mu = \left(\frac{1}{p_3} + \frac{p_3-1}{2p_3}\right) < 1$. Hence, by the Banach's fixed point theorem T has a unique fixed point which is a nonoscillatory solution of the system (1.1) on $[a, b]$. Thus the proof of the theorem is complete. □

Theorem 3.3. *Assume that* (A_1), (A_2), (A_3) *hold and* $-1 < -p_5 \leq p(t) \leq 0$, $p_5 > 0$ *for* $t \in \mathbb{R}_+$. *Then every solution of (1.1) converges to zero as* $t \to \infty$ *if and only if* (A_{13}) *holds.*

Proof. Proceeding as in the proof of Theorem 3.1, one can prove that $x(t)$ is bounded when $w(t) > 0$ or $w(t) < 0$ for $t \geq t_2$. Consequently, $\lim_{t \to \infty} z(t)$ exists and hence $\lim_{t \to \infty} w(t)$ exists. Using the same type of argument as in the proof of Theorem 3.1, it is easy to show that $\liminf_{t \to \infty} x(t) = 0$ and by Lemma 3.1, $\lim_{t \to \infty} z(t) = 0$. Indeed,

$$0 = \lim_{t \to \infty} z(t) = \limsup_{t \to \infty}\left(x(t) + p(t)x(t-\tau)\right)$$
$$\geq \limsup_{t \to \infty} x(t) + \liminf_{t \to \infty}\left(-p_5 x(t-\tau)\right)$$
$$= (1 - p_5)\limsup_{t \to \infty} x(t)$$

implies that $\limsup_{t \to \infty} x(t) = 0$ $[\because 1 - p_5 > 0]$. The rest of the sufficient part follows from Theorem 3.1.

In order to prove that the condition (A_{13}) is necessary, we assume that (3.4) holds. Then there exist $t_1 > 0$ such that

$$\int_{t_1}^{\infty} \sum_{i=1}^{m} q_i(\eta) d\eta < \frac{1-p_5}{10H(1)}, \quad t \geq t_1$$

and $|F(t) - M| < \frac{1-p_5}{20}$ for $t \geq t_2$. For $t_3 > \max\{t_1, t_2\}$, let $Y = BC([t_3, \infty), \mathbb{R})$ be the space of all real valued bounded continuous functions defined on $[t_3, \infty)$. Clearly, Y is a Banach space with respect to supremum norm defined by

$$||y|| = \sup\{|y(t)| : t \geq t_3\}.$$

Let $L = \{y \in Y : y(t) \geq 0, t \geq t_3\}$. Then, Y is a partially ordered Banach space (see for e.g., [10], p. 30). For $u, v \in X$, we define $u \leq v$ if and only if $u - v \in L$. Let

$$S = \left\{x \in Y : \frac{1-p_5}{20} \leq x(t) \leq 1, \ t \geq t_3\right\}.$$

If $x_0(t) = \frac{1-p_5}{20}$, then $x_0 \in S$ and $x_0 = g.l.b.\ S$. Further, if $\Phi \subset S^* \subset S$, then

$$S^* = \left\{x \in Y : l_1 \leq x(t) \leq l_2, \ \frac{1-p_5}{20} \leq l_1, \ l_2 \leq 1\right\}.$$

Let $v_0(t) = l'_2$, $t \geq t_3$, where $l'_2 = \sup\{l_2 : \frac{1-p_5}{20} \leq l_2 \leq 1\}$. Then $v_0 \in S$ and $v_0 = l.u.b.\ S^*$. For $t_4 = t_3 + \rho$, define $T : S \to S$ by

$$Tx(t) = \begin{cases} Tx(t_4), & t \in [t_3, t_4] \\ -p(t)x(t-\tau) + \frac{1-p_5}{10} + (F(t) - M) + \int_t^{\infty} \sum_{i=1}^{m} q_i(\eta) H(x(\eta - \sigma_i)) d\eta, & t \geq t_4. \end{cases}$$

For every $x \in S$,

$$Tx(t) \leq p_5 + H(1)\left[\int_t^{\infty} \sum_{i=1}^{m} q_i(\eta) d\eta\right] + \frac{1-p_5}{20} + \frac{1-p_5}{10} < \frac{1+3p_5}{4} < 1$$

and

$$Tx(t) \geq \frac{1-p_5}{10} + (F(t) - M) > \frac{1-p_5}{10} - \frac{1-p_5}{20} = \frac{1-p_5}{20}$$

implies that $Tx \in S$. Now, for $x_1, x_2 \in S$, it is easy to verify that $x_1 \leq x_2$ implies that $Tx_1 \leq Tx_2$. Hence by Knaster-Tarski fixed point theorem (see for e.g., [10], Theorem 1.7.3), T has a unique fixed point such that $\lim_{t \to \infty} x(t) \neq 0$. This completes the proof of the theorem. □

Theorem 3.4. *Assume that* (A_1), (A_2), (A_3) *hold and* $-\infty < -p_6 \leq p(t) \leq -p_7 < -1$ *for* $t \in \mathbb{R}_+$. *Let* H *be Lipschitzian on intervals of the form* $[\alpha, \beta]$, $0 < \alpha < \beta < \infty$. *Then every bounded solutions of* (1.1) *converges to zero as* $t \to \infty$ *if and only if* (A_{13}) *holds.*

Proof. Proceeding as in the proof of the Theorem 3.3 we have obtained $\lim_{t \to \infty} z(t) = 0$. Indeed,

$$0 = \lim_{t \to \infty} z(t) = \liminf_{t \to \infty} \big(x(t) + p(t)x(t-\tau)\big)$$
$$\leq \limsup_{t \to \infty} x(t) + \liminf_{t \to \infty} \big(-p_7 x(t-\tau)\big)$$
$$= (1 - p_7) \limsup_{t \to \infty} x(t)$$

implies that $\limsup_{t \to \infty} x(t) = 0$ [$\because 1 - p_7 < 0$]. Thus $\liminf_{t \to \infty} x(t) = 0$ and hence $\lim_{t \to \infty} x(t) = 0$. The case $x(t) < 0$ is similar.

For necessary part, we need to mention the following:

$$\int_{t_1}^{\infty} \sum_{i=1}^{m} q_i(\eta) d\eta < \frac{p_7 - 1}{2K}, \quad |F(t) - M| > \frac{1}{2}(p_7 - 1),$$

where $K = \max\{K_1, K_2\}$, K_1 is the Lipschitz constant of H on $[a, b]$, $K_2 = H(b)$ such that

$$a = \frac{\mu p_7 - p_6(p_7 - 1)}{p_6 p_7}, \quad b = \frac{1}{2} + \frac{\mu}{p_7 - 1}$$

for

$$\mu > \frac{p_6(p_7 - 1)}{p_7} > 0.$$

We set

$$S = \big\{x \in Y : a \leq x(t) \leq b, \, t \geq t_0\big\}$$

and

$$Tx(t) = \begin{cases} Tx(t_2 + \rho), & t \in [t_2, t_2 + \rho] \\ -\frac{x(t+\tau)}{p(t+\tau)} + \frac{F(t+\tau)}{p(t+\tau)} - \frac{\mu}{p(t+\tau)} + \frac{1}{p(t+\tau)} \Big[\int_{t+\tau}^{\infty} \sum_{i=1}^{m} q_i(\eta) H\big(x(\eta - \sigma_i)\big) d\eta \Big], & t \geq t_2 + \rho. \end{cases}$$

Therefore T is a contraction with a contraction constant $\frac{1+p_6}{2p_6}$. This completes the proof of the theorem. □

Remark 3.2. *In the above theorems, H could be linear, sublinear or superlinear.*

Remark 3.3. *Theorems 3.1 - 3.4 are hold true for $M = 0$.*

4. DISCUSSION AND EXAMPLES

All results of this chapter can hold for any type of first-order functional differential equations.

The chapters concludes with the following example to illustrate our main results.

Example 4.1. *Consider the neutral differential equation*

$$\bigl(x(t) + x(t - \pi)\bigr)' + x(t - 2\pi) + x(t - 4\pi) = 2\sin(t), \qquad (4.1)$$

where $p(t) = 1$, $q_1(t) = q_2(t) = 1$, $\tau = \pi$, $m = 2$, $\sigma_1 = 2\pi$, $\sigma_2 = 4\pi$, $H(x) = x$ *and* $f(t) = 2\sin(t)$. *Indeed, if we choose* $F(t) = -2\cos(t)$ *then* $F'(t) = f(t)$. *Since*

$$F^+(t) = \begin{cases} -2\cos(t), & 2n\pi + \frac{\pi}{2} \leq t \leq 2n\pi + \frac{3\pi}{2} \\ 0, & \text{otherwise,} \end{cases}$$

and

$$F^-(t) = \begin{cases} 2\cos(t), & 2n\pi + \frac{3\pi}{2} \leq t \leq 2n\pi + \frac{5\pi}{2} \\ 0, & \text{otherwise.} \end{cases}$$

Therefore

$$F^+(t - 2\pi) = \begin{cases} -2\cos(t), & 2n\pi + \frac{5\pi}{2} \leq t \leq 2n\pi + \frac{7\pi}{2} \\ 0, & \text{otherwise,} \end{cases}$$

and

$$F^-(t - 2\pi) = \begin{cases} 2\cos(t), & 2n\pi + \frac{7\pi}{2} \leq t \leq 2n\pi + \frac{9\pi}{2} \\ 0, & \text{otherwise.} \end{cases}$$

Also

$$F^+(t - 4\pi) = \begin{cases} -2\cos(t), & 2n\pi + \frac{9\pi}{2} \leq t \leq 2n\pi + \frac{11\pi}{2} \\ 0, & \text{otherwise,} \end{cases}$$

and

$$F^-(t - 4\pi) = \begin{cases} 2\cos(t), & 2n\pi + \frac{11\pi}{2} \leq t \leq 2n\pi + \frac{13\pi}{2} \\ 0, & \text{otherwise.} \end{cases}$$

Now

$$\int_{4\pi}^{\infty} [Q_1(\eta)F^+(\eta - 2\pi) + Q_2(\eta)F^+(\eta - 4\pi)]d\eta = I_1 + I_2,$$

where for $n = 0, 1, 2, \cdots$, we get

$$I_1 = \int_{4\pi}^{\infty} F^+(\eta - 2\pi)d\eta = \sum_{n=0}^{\infty} \int_{2n\pi + \frac{5\pi}{2}}^{2n\pi + \frac{7\pi}{2}} [-2\cos(\eta)]d\eta$$

$$= -2\sum_{n=0}^{\infty} \left[\sin(\eta)\right]_{2n\pi + \frac{5\pi}{2}}^{2n\pi + \frac{7\pi}{2}} = +\infty,$$

$$I_2 = \int_{4\pi}^{\infty} F^+(\eta - 4\pi)d\eta = \sum_{n=0}^{\infty} \int_{2n\pi + \frac{9\pi}{2}}^{2n\pi + \frac{11\pi}{2}} [-2\cos(\eta)]d\eta$$

$$= -2\sum_{n=0}^{\infty} \left[\sin(\eta)\right]_{2n\pi + \frac{9\pi}{2}}^{2n\pi + \frac{11\pi}{2}} = +\infty.$$

Clearly, (A_1), (A_2), (A_4), (A_5) and (A_7) *are satisfied. Hence, by Theorem 2.2, every solution of (4.1) is oscillatory. Thus, in particular,* $x(t) = sin(t)$ *is an oscillatory solution of the equation 4.1.*

Example 4.2. Consider the neutral differential equation

$$\big(x(t) - x(t - 2\pi)\big)' + x(t - 2\pi) + x(t - 4\pi) = 2sin(t), \quad (4.2)$$

where $p(t) = -1$, $q_1(t) = q_2(t) = 1$, $\tau = 2\pi$, $m = 2$, $\sigma_1 = 2\pi$, $\sigma_2 = 4\pi$, $H(x) = x$ and $f(t) = 2sin(t)$. Indeed, if we choose $F(t) = -2cos(t)$ then $F'(t) = f(t)$. Clearly, (A_2), (A_4), (A_5), (A_8) and (A_9) are hold true. Hence, Theorem 2.3 can be applied to (4.2), that is, every solution of (4.2) oscillates. Indeed, $x(t) = sin(t)$ is such a solution of (4.2).

In Section 3, Lemma 3.1 doesn't include $p(t) \equiv 1$ for all t (see for e.g., [10]). The present analysis doesn't allow the case $p(t) \equiv -1$ for all t. Hence in this discussion, a necessary and sufficient condition is established excluding $p(t) = \pm 1$ for all t. It seems that a different approach is necessary to study necessary and sufficient conditions for the case $p(t) = \pm 1$.

Example 4.3. *Consider the neutral differential equation*

$$\left(x(t) + e^{-\pi}x(t-\pi)\right)' + e^{3t-7\pi}\left(x(t-2\pi)\right)^3 = -e^{-t} \tag{4.3}$$

where $H(x) = x^3$. *Indeed, if we choose* $F(t) = e^{-t}$, *then* $F'(t) = f(t)$. *Clearly, conditions all assumptions of Theorem 3.1 hold true. Hence by Theorem 3.1 every solution of* (4.3) *converges to zero as* $t \to \infty$. *In particular,* $x(t) = e^{-t}$ *is such a solution of* (4.3).

REFERENCES

[1] Ahmed F. N., Ahmad R. R., Din U. K. S. and Noorani M. S. M.; Oscillation criteria of first order neutral delay differential equations with Variable Coefficients, *Abst. Appl. Anal.*, Vol. 2013, Article ID 489804, http://dx.doi.org/10.1155/2013/489804.

[2] Das P. and Misra N.; A necessary and sufficient condtion for the solutions of a functional differential equation to be oscillatory or tend to zero, *J. Math. Anal. Appl.*, 204 (1997), 78-87.

[3] Elabbasy E. M., Hassan T. S. and Saker S. H.; Oscillation criteria for first-order nonlinear neutral delay differential equations, *Electronic Journal of Differential Equations*, 2005 (2005), 134, 1-18.

[4] Graef J. R., Savithri R. and Thandapani E.; Oscillation of First Order Neutral Delay Differential Equations, *Electronic journal of qualitative theory of differential equations*, Proc. 7th Coll. QTDE, (2004), No. 12 1-11.

[5] Kubiaczyk I., Saker S. H. and J. Morchalo; New oscillation criteria for first order nonlinear neutral delay differential equations, *Appl. Math. Compu.*, 142 (2003), 225-242.

[6] Zhang W. P., Feng W., Yan J. R. and Song J. S.; Existence of Nonoscillatory Solutions of First-Order Linear Neutral Delay Differential Equations, *Computers and Mathematics with Applications*, 49 (2005), 1021-1027.

[7] Ahmed F. N., Ahmad R. R., Din U. K. S. and Noorani M. S. M.; Oscillation criteria for nonlinear functional differential equations of neutral type, *J. Inequ. Appl.*, DOI 10.1186/s13660-015-0608-5 (2015), 2015:97.

[8] Parhi N. and Rath R. N.; On oscillation and asymptotic behaviour of solutions of forced first order neutral differential equations, *Indian Acad. Sci, (Math. Sci.)*, Vol. III, No. 3, Aug (2001), 337-350.

[9] Parhi N. and Rath R. N.; Oscillation criteria for forced first order neutral differential equations with variable coefficients, *J. Math. Anal. Appl.*, 256 (2001), 525-541.

[10] Gyori I. and Ladas G.; *Oscillation Theory of Delay Differential Equations with Applications,* Oxford University press., (1991).

Bibliography

Ahmed F. N., Ahmad R. R., Din U. K. S. and Noorani Mohd. S. M.; Oscillations for Nonlinear Neutral Delay Differential Equations with Variable Coefficients, *Abstract and Applied Analysis*, Volume 2014, Article ID 179195, http://dx.doi.org/10.1155/2014/179195.

Berezansky L. and Braverman E.; Oscillation criteria for a linear neutral differential equation, *J. Math. Anal. Appl.* 286 (2003), 601-617.

Erbe L. H., Kong Q. and Zhang B. G.; *Oscillation Theory for Functional-Differential Equations.* Marcel Dekker, Inc., New York, (1995).

Kubiaezyk I. and Saker S. H.; Oscillation of solutions to neutral delay differential equations, *Math. Slovaca.*, 52 (2002), 3, 343-359.

Karpuz B. and Ocalan O.; *Oscillation criteria for some classes of linear delay differential equations of first-order,* Bulletin of the Institute of Mathematics, Academia Sinica (New Series)., 3(2): (2008), 293-314.

Karpuz B. and Santra S. S.; Oscillation theorems for second-order nonlinear delay differential equations of neutral type, *Hacet. J. Math. Stat.*, 48(3): (2019), 633-643.

Liu B. and Huang L.; Existence and uniqueness of periodic solutions for a kind of first order neutral functional differential equations, *J. Math. Anal. Appl.*, 322 (2006), 121-132.

Liu Z., Kangb S. M. and Ume J. S.; Existence of bounded nonoscillatory solutions of first-order nonlinear neutral delay differential equations, *Computers and Mathematics with Applications*, 59 (2010), 3535-3547.

Liu G. and Yan J.; Global asymptotic stability of nonlinear neutral differential equation, *Commun Nonlinear Sci Numer Simulat.*, 19 (2014) 1035-1041.

El-Morshedy H. A.; On the distribution of zeros of solutions of first order delay differential equations, *Nonlinear Analysis: Theory, Methods and Applications*, 74 (2011), 3353-3362.

Pinelas S. and Santra S. S.; Necessary and sufficient condition for oscillation of nonlinear neutral first-order differential equations with several delays, *J. Fixed Point Theory Appl.*, 20(27): (2018), 1-13.

Pinelas S. and Santra S. S.; Necessary and Sufficient Conditions for Oscillation of Nonlinear First Order Forced Differential Equations with Several Delays of Neutral Type, *Analysis,* 39(3): (2019), 97-105.

Santra S. S.; Oscillation criteria for nonlinear neutral differential equations of first order with several delays, *Mathematica*, Tome 57 (80)(1-2): (2015), 75-89.

Santra S. S.; Existence of positive solution and new oscillation criteria for nonlinear first-order neutral delay differential equations, *Differ. Equ. Appl.*, 8(1): (2016), 33-51.

S. S. Santra; Necessary and sufficient condition for oscillation of nonlinear neutral first order differential equations with several delays, *Mathematica*, 58 (81)(1-2) (2016), 85-94.

Santra S. S.; Oscillation analysis for nonlinear neutral differential equations of second-order with several delays, *Mathematica*, 59(82)(1-2): (2017), 111-123.

Santra S. S.; Oscillation analysis for nonlinear neutral differential equations of second-order with several delays and forcing term, *Mathematica*, 61(84)(1): (2019), 63-78.

Santra S. S.; Necessary and sufficient condition for the solutions of first-order neutral differential equations to be oscillatory or tend to zero, *KYUNGPOOK Math. J.*, 59 (2019), 73-82.

Santra S. S.; Necessary and sufficient condition for oscillatory and asymptotic behaviour of second-order functional differential equations, *Krag. J. Math.*, 44(3): (2020), 459-473.

Tang X. H.; Oscillation for first-order nonlinear delay differential equations, *J. Math. Anal. Appl.*, 264 (2001), 510-521.

Tang X. H. and Yu J. S.; Linearized oscillation of first-order nonlinear neutral delay differential equations, *J. Math. Anal. Appl.*, 258 (2001), 194-208.

Tang X. H. and Lin X.; Necessary and sufficient conditions for oscillation of first order nonlinear neutral differential equations, *J. Math. Anal. Appl.*, 321 (2006), 553-568.

Zhang X. and Yan J.; Oscillation criteria for first order neutral differential equations with positive and negative coefficients, *J. Math. Anal. Appl.*, 253 (2001), 204-214.

Chapter 5

OSCILLATION CRITERIA FOR NEUTRAL DIFFERENCE EQUATIONS

Shyam Sundar Santra[*], *Debasish Majumder*[†],
Rupak Bhattacharjee[‡] *and Tanusri Ghosh*[§]
Department of Mathematics, JIS College of Engineering,
Kalyani, India

Abstract

In this chapter, we present some sufficient conditions for the oscillation of solutions to the first and second-order neutral delay difference equation. Also, we state some open problems and some examples are presented to show effectiveness of the main results.

Keywords: oscillation, nonoscillation, difference equation, nonlinear, delay

Mathematics Subject Classification (2016): 39A10, 39A12

1. INTRODUCTION

Difference equation is getting an increasing importance among the practitioners in the recent past, since it has been identified as the effective tool to study

---

[*]Corresponding Author's Email: shyam01.math@gmail.com.
[†]Email: debamath@rediffmail.com.
[‡]Email: rupakb13@yahoo.co.in.
[§]Email: tanusrighosh1994@gmail.com.

the behaviour of discrete systems which, unlike continuous systems, cannot be studied using differential equation. Moreover, difference equation plays an important role in order to solve differential equation by finite difference method. Differential equations can be approximated by the use of difference equations which helps to develop high-speed digital computing machine. During the last one and half decade, a multidimensional application of difference equation has been observed in the field of statistics, science and engineering.

Besides the approximations of ordinary and partial differential equations, difference equation has emerged as a powerful tool for the analysis of electrical, mechanical, thermal, and other systems which involves the recurrence of identical sections. The study of the behaviour of the electric-wave filters, multistage amplifiers, magnetic amplifiers, insulator strings, continuous beams of equal span, crankshafts of multicylinder engines, acoustical filters, etc., are generally very lengthy, especially, when the large number of elements involved. The use of difference equation greatly reduces the complexity and expedites the study of such systems.

Courant, Friedrichs, and Lewy [1], in a celebrated paper in 1928, were the first to propose the use of difference equations for solving partial differential equations. However, the application of the proposed method was first noted around 1943 as a result of the stimulus of war-time technology and with the assistance of the first digital automatic computers [2]. Since then numerous problems involving time-dependent fluid flows, neutron diffusion and transport, radiation flow, thermo-nuclear reactions, and problems involving the solution of several simultaneous partial differential equations have been solved by the use of difference equations. In [3], Tanaka discussed the various solution of oscillation first-order neutral delay differential equations. Ocalan et al. [4] have carried out an extensive study on the problems of oscillation of neutral differential equation with positive and negative coefficients. In [5], Ayyappan et al. have considered a second order neutral delay difference equation of the form

$$\Delta\big(a_n \Delta^{m-1}(x_n + p_n x_{n-k}^\alpha)\big) + q_n x_{n-l}^\beta = 0 \qquad (1.1)$$

and obtained some new oscillation results for the equation (1.1) when $0 \leq p_n < 1$. Over the years, several studies have been carried out on oscillation of second order neutral difference equations [1, 6–13].

In the chapter, an attempt is made to study oscillation criteria of the neutral difference equation of the form

$$\Delta[x(n) + p(n)x(n-m)] + q(n)f\big(x(n-l)\big) = 0, \qquad (1.2)$$

where $f \in C(\mathbb{R}, \mathbb{R})$ satisfying the properties $xf(x) > 0$ for $x \neq 0$ for different ranges of the neutral coefficient p. Also, we have taken the second order neutral difference equation of the form

$$\Delta\left(r(n)(\Delta z(n))^\gamma\right) + q(n)x^\alpha(\sigma(n)) = 0, \qquad (1.3)$$

for $n \geq n_0$ where

$$z(n) = x(n) + p(n)x(\tau(n)), \qquad (1.4)$$

and $\Delta x(n) = x(n+1) - x(n)$, γ and α are the quotient of two odd positive integers, and p, q, r, σ, τ are real valued functions.

2. SUFFICIENT CONDITIONS FOR OSCILLATION OF (1.2)

Given a sequence of initial values $\{\phi(n)\}_{n \leq n_0}$, by a solution we mean sequence of real numbers $\{x(n)\}_{n=n_0}^\infty$ that satisfies (1.2) and agrees with the initial values.

A solution x is called oscillatory if $x(n) > 0$ for infinitely many indices n, and $x(n) < 0$ for infinitely many indices n. A solution x is called eventually positive if $x(n) > 0$ for all n large enough; similarly, x is eventually negative is $x(n) < 0$ for all n large enough.

In this section, we discuss the oscillation properties of solutions of the solutions of equation (1.2). Throughout our discussion we use the following notation:

$$y(n) = x(n) + p(n)x(n-m), \qquad (2.1)$$

Theorem 2.1. Let $-\infty < -r \leq p(n) \leq -1$, $r > 0$. Assume that $m \geq l$ hold. Furthermore, assume that

(A1) f is odd and $f(uv) = f(u)f(v)$, $u, v \in \mathbb{R}$,

(A2) $\int_{\pm c}^{\pm \infty} \frac{dx}{f(x)} < \infty$, $c > 0$;

and

(A3) $\sum_{n=1}^\infty q(n) = \infty$

hold. Then (1.2) *is oscillatory.*

Proof. On the contrary, let $x(n)$ be a solution of (1.2) such that $x(n) > 0$ or $x(n) < 0$ for $n \geq n_0$. Without loss of generality and due to $(A1)$, we may assume that $x(n), x(n-m), x(n-l) > 0$ for $n \geq n_1 = n_0 + \rho$. Using (2.1) in (1.2), we obtain

$$\Delta y(n) = -q(n)f(x(n-l)) \leq 0, \qquad (2.2)$$

for $n \geq n_1$. So, there exists $n_2 > n_1$ such that $y(n)$ is monotonic decreasing for $n \geq n_2$. We assert that $y(n) < 0$ for $n \geq n_2$. If not, let there exist $n_3 > n_2$ such that $y(n) \geq 0$ for $n \geq n_3$. As a result,

$$x(n) \geq -p(n)x(n-m) \geq x(n-m) \geq x(n-2m) \geq \cdots \geq x(n_3)$$

implies that $x(n)$ is bounded below by a constant (say) β. Summing (2.2) from n_3 to $n-1$, we obtain

$$\sum_{\eta=n_3}^{n-1} \Delta y(\eta) + \sum_{\eta=n_3}^{n-1} q(\eta)f(x(\eta-l)) = 0,$$

that is,

$$y(n) - y(n_3) + \sum_{\eta=n_3}^{n-1} q(\eta)f(x(\eta-l)) = 0.$$

Therefore,

$$y(n) = y(n_3) - \sum_{\eta=n_3}^{n-1} q(\eta)f(x(\eta-l))$$

implies that

$$y(n) \leq y(n_3) - f(\beta) \sum_{\eta=n_3}^{n-1} q(\eta) \quad \to -\infty \quad \text{as} \quad n \to \infty$$

a contradiction to the fact that $y(n) > 0$ for $n \geq n_3$. Hence, $y(n) < 0$ for $n \geq n_2$. Consequently, we can find a $n_3 > n_2$ such that

$$y(n) > p(n)x(n-m) \geq -rx(n-m)$$

implies that $y(n+m-l) \geq -rx(n-l)$ for $n \geq n_3$. Hence, (1.2) becomes

$$\Delta y(n) + \frac{q(n)}{f(-r)} f\big(y(n+m-l)\big) \leq 0, \qquad (2.3)$$

Since y is nonincreasing for $n \geq n_3$, then it follows that

$$\Delta y(n) + \frac{q(n)}{f(-r)} f\big(y(n)\big) \leq 0,$$

that is,

$$\frac{\Delta y(n)}{f\big(y(n)\big)} + \frac{q(n)}{f(-r)} \geq 0,$$

If $y(n+1) \leq u \leq y(n)$, then the preceding inequalities reduce to

$$\int_{y(n)}^{y(n+1)} \frac{du}{f(u)} + \frac{q(n)}{f(-r)} \geq 0.$$

Therefore,

$$\sum_{\eta=n_3}^{n} q(\eta) \leq -f(-r) \sum_{\eta=n_3}^{n} \int_{y(\eta)}^{y(\eta+1)} \frac{du}{f(u)} = -f(-r) \int_{y(n_3)}^{y(n+1)} \frac{du}{f(u)} < \infty$$

a contradiction to $(A3)$ due to $(A2)$. This completes the proof of the theorem. $\square$

Theorem 2.2. *Assume that all conditions of Theorem 2.1 hold except* $(A2)$. *Then every bounded solution of* (1.2) *oscillates.*

Proof. Proceeding as in the proof of Theorem 2.1, we obtain that $y(n) < 0$ for $n \geq n_2$. So, we can find a $n_3 > n_2$ and $\alpha > 0$ such that $y(n) \leq -\alpha$ for $n \geq n_3$. Consequently, (2.3) becomes

$$\Delta y(n) + \frac{f(\alpha)}{f(r)} q(n) \leq 0, \qquad (2.4)$$

for $n \geq n_3$. Summing (2.4) from n_3 to $n-1$, we get

$$y(n) - y(n_3) + \frac{f(\alpha)}{f(r)} \sum_{\eta=n_3}^{n-1} q(\eta) \leq 0,$$

that is,

$$\frac{f(\alpha)}{f(r)} \sum_{\eta=n_3}^{n-1} q(\eta) \leq y(n_3) - y(n) < \infty \quad \text{as} \quad n \to \infty,$$

a contradiction to $(A3)$. Hence, the theorem is proved. □

Theorem 2.3. *Let $-1 \leq -r \leq p(n) \leq 0$, $r > 0$. Assume that $(A1)$ and $(A3)$ hold. Furthermore, assume that*

$(A4)$ $\int_0^{\pm c} \frac{dx}{f(x)} < \infty$, $0 < c < \infty$

hold. Then every solution of (1.2) oscillates.

Proof. Proceeding as in Theorem 2.1, we obtain that $y(n)$ is monotonic decreasing for $n \geq n_2$. So, there exists $n_3 > n_2$ such that $y(n) > 0$ or < 0 for $n \geq n_3$. Assume that $y(n) > 0$ for $n \geq n_3$. Then $y(n) \leq x(n)$ for $n \geq n_3$. Consequently, (2.2) reduce to

$$\Delta y(n) \leq -q(n) f\big(y(n-l)\big), \tag{2.5}$$

for $n \geq n_4 > n_3 + l$ and due to decreasing $y(n)$,

$$\frac{\Delta y(n)}{f\big(y(n)\big)} \leq -q(n).$$

Since $\lim_{n \to \infty} y(n) < \infty$, then proceeding as in Theorem 2.1, we obtain a contradiction to $(A3)$. Indeed,

$$\sum_{\eta=n_4}^{n-1} q(\eta) \leq -\sum_{\eta=n_4}^{n-1} \frac{\Delta y(\eta)}{f(y(\eta))} \leq -\sum_{\eta=n_4}^{n-1} \int_{y(\eta)}^{y(\eta+1)} \frac{du}{f(u)} = -\int_{y(n_4)}^{y(n)} \frac{du}{f(u)}$$

where $y(\eta+1) < x < y(\eta)$. Hence, $y(n) < 0$ for $n \geq n_3$. From (2.1), it follows that

$$x(n) < -p(n)x(n-m) \leq x(n-m) \leq x(n-2m) \leq \cdots \leq x(n_3).$$

Indeed, the above observation reveals that $x(n)$ is bounded for $n \geq n_3$. The rest of the proof follows from Theorem 2.2. Hence, the proof of the theorem is completed. □

Theorem 2.4. Let $-1 \leq -r \leq p(n) \leq 0$, $r > 0$. Assume that $(A1)$ and $(A3)$ hold. If

$(A5)$ there exists $\mu > 0$ such that $|f(u)| \geq \mu|u|$, $u \in \mathbb{R}$

and

$(A6)$ $\limsup_{n \to \infty} \sum_{\eta=n_4}^{n-1} q(\eta) > \frac{1}{\mu}$, $l \geq 1$

hold, then (1.2) is oscillatory.

Proof. Let $x(n)$ be a nonoscillatory solution of (1.2) such that $x(n) > 0$, $x(n-m) > 0$, $x(n-l) > 0$ for $n \geq n_1 = n_0 + l$. Proceeding as in Theorem 2.3, we get a contradiction to $(A3)$ when $y(n) < 0$ for $n \geq n_3$. Assume that $y(n) > 0$ for $n \geq n_3$. Therefore, (2.5) holds for $n \geq n_4 = n_3 + l$. Summing (2.5) from n_4 to $n-1$, we obtain

$$y(n_4) - y(n-1) + \sum_{\eta=n_4}^{n-1} q(\eta) f\big(y(\eta-l)\big) \leq 0,$$

that is,

$$-yn(n_4) + \sum_{\eta=n_4}^{n-1} q(\eta) f\big(y(\eta-l)\big) \leq 0.$$

Using the fact that z is decreasing, the last inequality yields

$$-y(n_4) + \mu y(n_4) \sum_{\eta=n_4}^{n-1} q(\eta) \leq 0,$$

due to $(A5)$. Consequently,

$$\limsup_{n \to \infty} \sum_{\eta=n_4}^{n-1} q(\eta) \leq \frac{1}{\mu}$$

contradicts $(A6)$. Thus, the proof of the theorem is completed. $\square$

Theorem 2.5. Let $p(n) \leq -1$ and $m - l > 0$. Assume that $(A1)$, $(A3)$ and $(A5)$ hold. For $m - l > 1$, if

$(A7)$ $\limsup_{n \to \infty} \sum_{\eta=n+l-m}^{n-1} \frac{-q(\eta)}{r(\eta+m-l)} > \frac{1}{\mu}$

holds, then (1.2) is oscillatory.

Proof. Proceeding as in Theorem 2.1 we have a contradiction to $(A3)$ when $y(n) > 0$ for $n \geq n_2$. Assume that $y(n) < 0$ for $n \geq n_2$. Consequently, there exists $n_3 > n_2$ such that

$$y(n) > p(n)x(n-m),$$

that is, $x(n-l) > \frac{y(n+m-l)}{r(n+m-l)}$ for $n \geq n_3$. Hence, (1.2) reduces to

$$\Delta y(n) + \mu q(n)\frac{y(n+m-l)}{r(n+m-l)} \leq 0, \tag{2.6}$$

due to $(A5)$. Summing (2.6) from $n+l-m$ to $n-1$ ($n \geq n_3 + m - l$), we have

$$y(n) - y(n+l-m) + \mu \sum_{\eta=n+l-m}^{n-1} q(\eta)\frac{y(\eta+m-l)}{r(\eta+m-l)} \leq 0,$$

that is,

$$y(n) + \mu \sum_{\eta=n+l-m}^{n-1} q(\eta)\frac{y(\eta+m-l)}{r(\eta+m-l)} \leq 0.$$

Since z s decreasing and $n+l-m \leq n-1$, $n+l-m \leq \eta$, then the preceding inequality becomes

$$y(n)\left[1 + \mu \sum_{\eta=n+l-m}^{n-1} \frac{q(\eta)}{r(\eta+m-l)}\right] \leq 0,$$

that is,

$$\sum_{\eta=n+l-m}^{n-1} \frac{-q(\eta)}{r(\eta+m-l)} \leq \frac{1}{\mu},$$

a contradiction to $(A7)$. Hence, the theorem is proved. $\square$

Theorem 2.6. *Let $0 \leq p(n) \leq r_1 < \infty$ for $m \leq l$. Assume that $(A1)$ hold. Furthermore, assume that*

$(A8)$ *there exists $\lambda > 0$ such that $f(u) + f(v) \geq \lambda f(u+v)$, $u, v \in \mathbb{R}_+$*

and

$(A9)$ $\sum_{n=m}^{\infty} Q(n) = \infty$,

hold, where $Q(n) = \min\{q(n), q(n-m)\}$, $n \geq m$. Then every solution of (1.2) oscillates.

Proof. On the contrary, we proceed as in Theorem 2.1 to obtain that $y(n)$ is decreasing for $n \geq n_2$. So, there exists $n_3 > n_2$ such that $y(n) > 0$ for $n \geq n_3$. It is easy to verify that

$$\Delta y(n) + q(n)f(y(n-l)) + f(r_1)\Delta y(n-m) + f(r_1)q(n-m)f(y(n-l-m)) \leq 0,$$

Applying $(A8)$ and $(A1)$ the last inequalities, we obtain

$$\Delta y(n) + f(r_1)\Delta y(n-m) + \lambda Q(n)f(y(n-l)) \leq 0$$

Using the fact that z is decreasing and $m \leq l$, we can find $n_4 > 0$ such that the above inequalities can be written as

$$\frac{\Delta y(n)}{f(y(n))} + f(r_1)\frac{\Delta y(n-m)}{f(y(n-m))} + \lambda Q(n) \leq 0, \qquad (2.7)$$

for $n \geq n_4$. If $y(n+1) \leq t \leq y(n)$, $y(n-m+1) \leq x \leq y(n-m)$ then form (2.7) it is easy to verify that

$$\int_{y(n)}^{y(n+1)} \frac{dt}{f(t)} + f(r_1)\int_{y(n-m)}^{y(n+1-m)} \frac{dx}{f(x)} + \lambda Q(n) \leq 0,$$

that is,

$$\sum_{\eta=n_4}^{n}\left[\int_{y(\eta)}^{y(\eta+1)} \frac{dt}{f(t)} + f(r_1)\int_{y(\eta-m)}^{y(\eta+1-m)} \frac{dx}{f(x)}\right] + \lambda \sum_{\eta=n_4}^{n} Q(\eta) \leq 0,$$

Consequently,

$$\lambda \sum_{\eta=n_4}^{\infty} Q(\eta) \leq -\lim_{n\to\infty}\left[\int_{y(n_4)}^{y(n+1)} \frac{dt}{f(t)} + f(r_1)\int_{y(n_4-m)}^{y(n+1-m)} \frac{dx}{f(x)}\right] < \infty$$

a contradiction to $(A9)$. This completes the proof of the theorem. □

Theorem 2.7. Let $0 \leq p(n) \leq r_1 \leq 1$ and $2m \leq l$. If $(A1)$, $(A5)$ and

$(A10)$ $\limsup_{n\to\infty} \sum_{\eta=n-m}^{n-1} Q(\eta) > \frac{1+r_1}{\mu}$

hold, then every solution of of (1.2) oscillates, where $Q(n)$ is defined in Theorem 2.6.

Proof. Proceeding as in Theorem 2.6, we obtain that $y(n) > 0$ and decreasing for $n \geq n_3$. Using $(A5)$ in (1.2), we get

$$\Delta y(n) + \mu q(n) x(n-l) \leq 0 \qquad (2.8)$$

due to (2.1). Upon using (2.8), we obtain

$$\Delta y(n) + \mu q(n) x(n-l) + r_1 \big[\Delta y(n-m) + \mu q(n-m) x(n-l-m)\big] \leq 0,$$

that is,

$$\Delta y(n) + r_1 \Delta y(n-m) + \mu Q(n) y(n-l) \leq 0, \qquad (2.9)$$

for $n \geq n_4 > n_3$. Summing (2.9) from $n-m$ to $n-1$, it follows that

$$y(n) - y(n-m) + r_1 y(n-m) - r_1 y(n-2m) + \mu \sum_{\eta=n-m}^{n-1} Q(\eta) y(\eta-l) \leq 0.$$

Therefore,

$$-y(n-m) - r_1 y(n-2m) + \mu \sum_{\eta=n-m}^{n-1} Q(\eta) y(\eta-l) \leq 0. \qquad (2.10)$$

Using the fact that y is decreasing and $\eta \leq n-1 < n$ in (2.10), we get

$$-y(n-m) - r_1 y(n-2m) + \mu y(n-l) \sum_{\eta=n-m}^{n-1} Q(\eta) \leq 0.$$

Hence,

$$y(n-2m) \Big[-1 - r_1 + \mu \sum_{\eta=n-m}^{n-1} Q(\eta) \Big] \leq 0$$

implies that

$$\sum_{\eta=n-m}^{n-1} Q(\eta) \leq \frac{1+r_1}{\mu}$$

which contradicts $(A10)$. This completes the proof. □

Example 2.1. *Consider the difference equation of the form:*

$$\Delta\left[x(n) - \frac{1}{e}x(n-2)\right] + (e^3-1)(e+1)e^{\frac{2n-7}{3}}y^{\frac{1}{3}}(n-2) = 0, \ n > 2, \tag{2.11}$$

where $m = 2 = l$, $p(n) = -\frac{1}{e}$, $q(n) = (e^3-1)(e+1)e^{\frac{2n-7}{3}}$, $f(u) = u^{\frac{1}{3}}$. *Since*

$$\sum_{n=1}^{\infty} q(n) = \sum_{n=1}^{\infty}(e^3-1)(e+1)e^{\frac{2n-7}{3}} = (e^3-1)(e+1)\sum_{n=1}^{\infty} e^{\frac{2n-7}{3}} = \infty,$$

then $(A3)$ *holds true. Indeed, all conditions of Theorem 2.3 also hold true. Hence, (2.11) is oscillatory. In particular,* $x(n) = (-1)^n e^n$ *is an oscillatory of (2.11).*

In the future scope related to this work the following points can be considered.

1. Necessary and sufficient conditions for the oscillation of (1.2) for the different ranges of the neutral coefficient $p(n)$.

3. SUFFICIENT CONDITIONS FOR OSCILLATION OF (1.3)

Given a sequence of initial values $\{\phi(n)\}_{n \leq n_0}$, by a solution we mean sequence of real numbers $\{x(n)\}_{n=n_0}^{\infty}$ that satisfies (1.3) and agrees with the initial values.

A solution x is called oscillatory if $x(n) > 0$ for infinitely many indices n, and $x(n) < 0$ for infinitely many indices n. A solution x is called eventually positive if $x(n) > 0$ for all n large enough; similarly, x is eventually negative is $x(n) < 0$ for all n large enough.

Note the restriction that γ and α be the quotient of odd integers can avoided by using $|x|^{\alpha-1}x$ instead of x^α, and the results below are still valid. Also note that if x is a solution of (1.3), then $-x$ is also a solution of (1.3). Therefore when a solution is of one sign we consider only the positive case.

In this section, we need the following assumption for our work in the sequel.

(A11) $\sigma(n) < n$, $\tau(n) < n$, $\lim_{n \to \infty} \sigma(n) = \infty$, $\lim_{n \to \infty} \tau(n) = \infty$.

(A12) $0 < p(n), 0 \leq q(n)$, for all $n \geq 0$.

(A13) $R(n) = \sum_{s=0}^{n-1} r^{-1/\gamma}(s) \to \infty$ as $n \to \infty$.

(A14) $-1 < -p_0 \leq p(n) \leq 0$ for $n \geq n_0$.

Lemma 3.1. *Assume that (A11)–(A14) and (1.4) are satisfied. If x is an eventually positive solution of (1.3), then there exists $n_1 > n_0$ such that*

$$\Delta\big(r(n)(\Delta z(n))^\gamma\big) \leq 0, \quad \forall n \geq n_1. \tag{3.1}$$

Furthermore, only one the following two cases happens.
Case 1:
$$\lim_{n \to \infty} x(n) = 0.$$

Case 2: there exist $n_5 > n_1$ and a positive constant δ such that for all $n \geq n_5$ we have:

$$0 < z(n) \tag{3.2}$$
$$0 < \Delta z(n) \tag{3.3}$$
$$z(n) \leq \delta R(n), \tag{3.4}$$
$$\big(R(n) - R(n_5 + 1)\big)\Big(\sum_{s=n}^{\infty} q(s) x^\alpha(\sigma(s))\Big)^{1/\gamma} \leq z(n). \tag{3.5}$$

Proof. Let x be an eventually positive solution. Then by (A11) there exists $n_1 \geq n_0$ such that $x(n) > 0$, $x(\tau(n)) > 0$ and $x(\sigma(n)) > 0$ for all $n \geq n_1$. From (1.3) it follows that

$$\Delta\big(r(n)(\Delta z(n))^\gamma\big) = -q(n) x^\alpha(\sigma(n)) \leq 0.$$

for $n \geq n_1$ which implies (3.1). Therefore, $r(n)(\Delta z(n))^\gamma$ is non-increasing for $n \geq n_1$. So, there exists $n_2 \geq n_1$ such that $r(n)(\Delta z(n))^\gamma$ is of only one sign. Since $r(n) > 0$ and γ is the quotient of two odd positive integers, it follows that $\Delta z(n)$ is also of only one sign for $n \geq n_2$. Therefore, $z(n)$ is monotonic on some interval $[n_3, \infty)$. Then there exist exists $n_4 \geq n_3$, so that z fits in only one of the following two cases:

Case 1: $z(n) \leq 0$ for all $n \geq n_4$. From the definition $z(n) = x(n) + p(n) x(\tau(n))$ and (A14), we have

$$0 < x(n) \leq -p(n) x(\tau(n)) \leq p_0 x(\tau(n)) \quad \forall n \geq n_4.$$

Since $\lim_{n\to\infty}\tau(n)=\infty$, there exists $b\geq 0$ such that $b=\limsup_{n\to\infty}x(n)=\limsup_{n\to\infty}x(\tau(n))$. Taking the limit superior in the above inequality, we have $0\leq b\leq p_0 b$. Since $0<p_0<1$ it follows that $0=b=\limsup_{n\to\infty}x(n)$. Then $x(n)$ begin positive, it converges to zero.

Case 2: $z(n)>0$ for all $n\geq n_4$. Recall that by (3.1), $r(n)(\Delta z(n))^\gamma$ is non-decreasing. Now we claim that $r(n)(\Delta z(n))^\gamma>0$. To obtain a contradiction assume that $r(n)(\Delta z(n))^\gamma\leq 0$ for some $n\geq n_4$. Then there exists $n_5\geq n_4$ such that $r(n)(\Delta z(n))^\gamma<0$ for all $n\geq n_5$. From $r(n)(\Delta z(n))^\gamma$ begin non-decreasing,

$$\Delta z(n)\leq \left(\frac{r(n_5)}{r(n)}\right)^{1/\gamma}\Delta z(n_5)\quad \text{for } n\geq n_5.$$

Summing from n_5 to $n-1$, we have

$$z(n)\leq z(n_5)+\big(r(n_5)\big)^{1/\gamma}\big(R(n)-R(n_5+1)\big)\Delta z(n_5). \qquad (3.6)$$

By (A13) the right-hand side approaches $-\infty$ as n approaches ∞, which contradicts that $z(n)>0$.

Therefore $r(n)(\Delta z(n))^\gamma>0$ for all $n\geq n_5$. From $r(n)(\Delta z(n))^\gamma>0$ and $r(n)>0$ we obtain (3.3).

Since $r(n)$ is the only variable in right-hand side of (3.6), and $\lim_{n\to\infty}R(n)=\infty$, there exists a positive constant δ such that (3.4) holds.

Because $r(n)(\Delta z(n))^\gamma$ is positive and non-increasing, $\lim_{n\to\infty}r(n)(\Delta z(n))^\gamma$ exists and is non-negative. Summing (1.3) from n to $a-1$, we obtain

$$r(a)(\Delta z(a))^\gamma-r(n)(\Delta z(n))^\gamma=-\sum_{s=n}^{a-1}q(s)x^\alpha(\sigma(s)).$$

Taking limit as $a\to\infty$, we have

$$r(n)(\Delta z(n))^\gamma\geq \sum_{s=n}^{\infty}q(s)x^\alpha(\sigma(s)) \qquad (3.7)$$

which implies

$$\Delta z(n)\geq \left(\frac{1}{r(n)}\sum_{s=n}^{\infty}q(s)x^\alpha(\sigma(s))\right)^{1/\gamma}.$$

Adding from n_5 to $n-1$, we have

$$z(n) \geq \sum_{k=n_5}^{n-1} \left(\frac{1}{r(k)} \sum_{s=k}^{\infty} q(s) x^\alpha(\sigma(s))\right)^{1/\gamma}.$$

For n and k satisfying $n_5 \leq k \leq n-1$, we have

$$z(n) \geq \sum_{k=n_5}^{n-1} \frac{1}{r^{1/\gamma}(k)} \left(\sum_{s=n}^{\infty} q(s) x^\alpha(\sigma(s))\right)^{1/\gamma}$$

$$= (R(n) - R(n_5+1)) \left(\sum_{s=n}^{\infty} q(s) x^\alpha(\sigma(s))\right)^{1/\gamma}$$

which is (3.5). This completes the proof. □

3.1. The Case $\alpha \leq \gamma$

In this subsection we assume that there exists a quotient of odd positive number β such that

$$\alpha \leq \beta \leq \gamma. \tag{3.8}$$

Theorem 3.1. *Assume* (A11)–(A14) *and* (3.8) *hold. Then every solution of* (1.3) *is either oscillatory or converges to zero, if*

$$\sum_{s=0}^{\infty} q(s) R^\alpha(\sigma(s)) = \infty. \tag{3.9}$$

Proof. We prove this by contradiction. Assume that (3.9) holds and x is an eventually positive which does not converge to zero. From Lemma 3.1, inequalities (3.2)–(3.5) are valid for $n \geq n_5$. Let

$$w(n) = \sum_{s=n}^{\infty} q(s) x^\alpha(\sigma(s)) \geq 0 \quad \text{for } n \geq n_5. \tag{3.10}$$

Then $\Delta w(n) = -q(n) x^\alpha(\sigma(n)) \leq 0$; thus w is non-increasing. From (3.5) and that $R(n) \to \infty$, there exists $n_6 \geq n_5$ such that

$$z(n) \geq \frac{1}{2} R(n) w^{1/\gamma}(n). \tag{3.11}$$

Using that $x \geq z$, (3.4), $\alpha - \beta \leq 0$, and the above inequality we have

$$\begin{aligned}x^\alpha(n) &\geq z^{\alpha-\beta}(n)z^\beta(n) \\ &\geq (\delta R(n))^{\alpha-\beta}\left(\frac{R(n)w^{1/\gamma}(n)}{2}\right)^\beta \\ &= \frac{\delta^{\alpha-\beta}}{2^\beta}R^\alpha(n)w^{\beta/\gamma}(n) \quad \text{for } n \geq n_6.\end{aligned}$$

Since $\sigma(n) \leq n$ and w is non-increasing,

$$\begin{aligned}x^\alpha(\sigma(n)) &\geq \frac{\delta^{\alpha-\beta}}{2^\beta}R^\alpha(\sigma(n))w^{\beta/\gamma}(\sigma(n)) \\ &\geq \frac{\delta^{\alpha-\beta}}{2^\beta}R^\alpha(\sigma(n))w^{\beta/\gamma}(n).\end{aligned} \quad (3.12)$$

Then

$$w^{-\beta/\gamma}(n) \geq \frac{\delta^{\alpha-\beta}}{2^\beta}R^\alpha(\sigma(n))x^{-\alpha}(\sigma(n)).$$

Therefore,

$$w^{-\beta/\gamma}(n) \geq \frac{1}{m2^\beta}\delta^{\alpha-\beta}R^\alpha(\sigma(n))x^{-\alpha}(\sigma(n)). \quad (3.13)$$

Next we consider the sequence $w^{1-\beta/\gamma}$ which is non-negative, and non-increasing. In fact we estimate $\Delta w^{1-\beta/\gamma}$ using a Taylor polynomial of order 1 for the function $h(x) = x^{1-\beta/\gamma}$. For $0 < \beta < \gamma$, about $x = a$ we have

$$b^{1-\beta/\gamma} - a^{1-\beta/\gamma} \leq \left(1 - \frac{\beta}{\gamma}\right)a^{-\beta/\gamma}(b-a).$$

So that

$$\Delta\left(w^{1-\beta/\gamma}(n)\right) \leq \left(1 - \frac{\beta}{\gamma}\right)w^{-\beta/\gamma}(n)\Delta w(n). \quad (3.14)$$

Summing (3.14) from n_6 to $n-1$ and then using that $w > 0$, we obtain

$$\begin{aligned}w^{1-\beta/\gamma}(n_6) &\geq \left(1 - \frac{\beta}{\gamma}\right)\left(-\sum_{s=n_2}^{n-1}w^{-\beta/\gamma}(s)\Delta w(s)\right) \\ &= \left(1 - \frac{\beta}{\gamma}\right)\sum_{s=n_2}^{n-1}w^{-\beta/\gamma}(s)\left(q(s)x^\alpha(\sigma(s))\right).\end{aligned} \quad (3.15)$$

Substituting (3.13) in the above inequality, and using that all summands in (3.13) and in (3.10) are non-negative, we have

$$w^{1-\beta/\gamma}(n_6) \geq \frac{\left(1-\frac{\beta}{\gamma}\right)}{m2^\beta} \sum_{s=n_2}^{n-1} q(s) x^\alpha(\sigma(s)) \delta^{\alpha-\beta} R^\alpha(\sigma(s)) x^{-\alpha}(\sigma(s))$$

$$= \frac{\left(1-\frac{\beta}{\gamma}\right)}{m2^\beta} \sum_{s=n_2}^{n-1} q(s) \delta^{\alpha-\beta} R^\alpha(\sigma(s)).$$

(3.16)

Note that (3.7) implies

$$\lim_{n\to\infty} \sum_{s=n_2}^{n-1} q(s) \delta^{\alpha-\beta} R^\alpha(\sigma(s)) = \infty,$$

which contradicts (3.16). This completes the proof of the theorem for eventually positive solutions. For eventually negative solutions we use the above process for $-x$ which is also a solution. □

3.2. General Case for α and γ

Lemma 3.2. *Assume* (A11)–(A14) *hold,* $\Delta p(n) \geq 0$, *and*

$$\sum_{s=n_0}^{\infty} q(s)(\sigma(s))^\alpha = \infty. \quad (3.17)$$

If x is an eventually positive solution of (1.3) *then either x converges to zero, or there exists n_6 such that*

$$\frac{\sigma(n)}{n} \leq \frac{z(\sigma(n))}{z(n)} \quad \text{for } n \geq n_6. \quad (3.18)$$

Proof. From Lemma 3.1, either $\lim_{n\to\infty} x(n) = 0$, or there exist n_5 such that (3.2)–(3.5) hold for $n \geq n_5$.

Now we claim that $\Delta(\Delta z(n))^\gamma$ and $\Delta(\Delta z(n))$ have the same sign. This follows from the mapping $t \mapsto t^\gamma$ being increasing for $t > 0$ and $\gamma > 0$. Using the product rule $\Delta(fg(n)) = f(n+1)\Delta g(n) + (\Delta f(n))g(n)$ and inequality (3.1) we have

$$r(n+1)\Delta\left((\Delta z(n))^\gamma\right) + \left(\Delta p(n)\right)(\Delta z(n))^\gamma \leq 0;$$

therefore $\Delta(\Delta z(n))^\gamma \leq 0$, and $\Delta(\Delta z(n)) \leq 0$. The same inequality is obtained when using the product rule $\Delta(fg(n)) = (\Delta f(n))g(n+1) + f(n)\Delta g(n)$.

Let us define
$$\phi(n) = z(n) - n\Delta z(n) \quad \text{for } n \geq n_5.$$

Then using $\Delta(fg(n)) = f(n+1)\Delta g(n) + (\Delta f(n))g(n)$, we have
$$\Delta\phi(n) = -(n+1)\Delta(\Delta z(n)) \geq 0,$$

so that ϕ is non-decreasing. To use the second version of the product rule we define $\psi(n) = z(n+1) - \Delta z(n)$.

Now we claim that ϕ is eventually non-negative. To obtain a contradiction, suppose that $\phi(n) < 0$ for all $n \geq n_5$. Then

$$\Delta\left(\frac{z(n)}{n}\right) = \frac{n\Delta z(n) - z(n)}{n(n+1)} = -\frac{\phi(n)}{n(n+1)}. \tag{3.19}$$

So that the above expression is positive, and $z(n)/n$ is strictly increasing. From (A11) there is $n^* \geq n_5$ such that $\sigma(n) \geq n_5$ for all $n \geq n^*$, in which case
$$\frac{z(\sigma(n))}{(\sigma(n))} \geq \frac{z(\sigma(n_5))}{\sigma(n_5)} > 0.$$

Summing (1.3) from n^* to $n-1$, using that $x \geq z$, and the above inequality, it follows that

$$r(n)(\Delta z(n))^\gamma - r(n^*)(\Delta z(n^*))^\gamma \leq -\sum_{s=n^*}^{n-1} q(s)z^\alpha(\sigma(s))$$

$$\leq -\sum_{s=n^*}^{n-1} q(s)z^\alpha(\sigma(s))$$

$$\leq -\sum_{s=n^*}^{n-1} q(s)\left(\frac{z(\sigma(n_5))}{\sigma(n_5)}\sigma(s)\right)^\alpha.$$

Therefore,
$$r(n^*)(\Delta z(n^*))^\gamma \geq \left(\frac{z(\sigma(n_5))}{\sigma(n_5)}\right)^\alpha \sum_{s=n^*}^{n-1} q(s)(\sigma(s))^\alpha$$

for $n > n^*$. By assumption (3.17), as $n \to \infty$, the right-hand side approaches ∞ while the left-hand side remains constant. This contradiction implies the existence of $n_6 \geq n_5$ such that $\phi(n) > 0$ for $n \geq n_6$. Then (3.19) implies $z(n)/n$ begin decreasing. Therefore $\sigma(n) \leq n$ implies (3.18) and completes the proof. □

Theorem 3.2. *Assume* (A11)–(A14) *hold, and* $\Delta p(n) \geq 0$. *Then every solution of* (1.3) *is either oscillatory or converges to zero, if for at least one index* i *we have*

$$\sum_{s=n_0}^{\infty} q(s)\bigl(\sigma(s)\bigr)^{\alpha} = \infty. \tag{3.20}$$

Proof. We prove this by contradiction. Assume that (3.20) holds (without lost of generality for $i = 1$) and x is an eventually positive solution not converging to zero. Then from Lemma 3.1, inequalities (3.2)–(3.5) hold for $n \geq n_5$.

As a motivation for introducing a Riccati transformation we do the following: The continuous version of (1.3) is divided by z^{α}, then add and subtract $(r/z^{\alpha})'(z')^{\gamma}$ to have product rule applied to a function w. For (1.3), define the Riccati transformation

$$w(n) = \frac{r(n)\bigl(\Delta z(n)\bigr)^{\gamma}}{z^{\alpha}(n)} \quad \text{for } n \geq n_5.$$

Then using that $x \geq z$ and $\Delta z(n) > 0$ (hence $\Delta z^{\alpha}(n) > 0$ because $z > 0$ and $\alpha > 0$), we have

$$\begin{aligned}
\Delta w(n) &= -\frac{q(n)x^{\alpha}\bigl(\sigma(n)\bigr)}{z^{\alpha}(n)} - r(n)\bigl(\Delta z(n)\bigr)^{\gamma}\frac{\Delta z^{\alpha}(n)}{z^{\alpha}(n+1)z^{\alpha}(n)} \\
&\leq -\frac{q(n)x^{\alpha}\bigl(\sigma(n)\bigr)}{z^{\alpha}(n)} \\
&\leq -q(n)\left(\frac{z\bigl(\sigma(n)\bigr)}{z(n)}\right)^{\alpha} \\
&\leq -q(n)\left(\frac{\sigma(n)}{n}\right)^{\alpha},
\end{aligned}$$

where we used (3.18). Summing from n_5 to $n-1$, we have

$$w(n) - w(n_5) \leq -\sum_{s=n_5}^{n-1} q(s)\left(\frac{\sigma(s)}{n}\right)^{\alpha};$$

thus
$$w(n_5) \geq \sum_{s=n_5}^{n-1} q(s)\left(\frac{\sigma(s)}{n}\right)^\alpha.$$

By (3.17), as $n \to \infty$, the right-hand side approaches ∞, while the left-hand side is constant. This contradiction completes the prof for the sufficiency part. □

In the future scope related to this work the following points can be considered.

1. Sufficient conditions for the oscillation of (1.3) for the other ranges of the neutral coefficient $p(n)$.

2. Necessary and sufficient conditions for the oscillation of (1.3) for the different ranges of the neutral coefficient $p(n)$.

REFERENCES

[1] Courant R., Friedrichs K., Lewyt H.; *On the Partial Difference Equations of Mathematical Physics,* Mathematische Annalen, **100** (1928), 32–74.

[2] Pipes L. A.; *Difference Equations and Their Applications,* Mathematics Magazine, **32**(5): (1959), 231–246.

[3] Tanaka S. *Oscillation of solution of first order neutral delay differential equations,* Hiroshima Math. J., **32** (2002), 73–85.

[4] Ocalan O.; *Oscillation of neutral differential equation with positive and negative coefficients,* J. Math. Anal. Appl. **331**(1): (2007), 644–654.

[5] Ayyappan G., Nithyakala G.; *Some oscillation results for even order delay difference equations with a sublinear neutral term,* Abst. Appl. Anal., **2018**, Article ID 2590158.

[6] Aktas M. F., Tiryaki A., Zafer A.; *Oscillation of third-order nonlinear delay difference equations,* Turk J Math., **36** (2012), 422–436.

[7] Han Z., Li T., Sun S., Chen W.; *Oscillation of second order quasilinear neutral delay differential equations* J. Appl. Math. Comput., **40** (2012), 1432-152.

[8] Jawahar G. G.; *Oscillaton behaviour of second order neutral delay difference equations,* Int. J. Mecha. Eng. Tech., bf 10(2): (2019), 150–154.

[9] Li H. J., Yeh C. C.; *Oscillation criteria for second-order neutral delay difference equations,* Comput. Math. Appl., **36**(10-12): (1998), 123–132.

[10] Parhi N., Tripathy A. K.; *Oscillation of a class of neutral difference equations of first-order,* J. Differ. Equ. Appl., **9**(10); (2003), 933–946.

[11] Parhi N., Tripathy A. K.; *Oscillation of forced nonlinear neutral deley difference equations of first-order,* Czechoslovak J. Math., **53**(1), (2003), 83–101.

[12] Saker S. H.; *New oscillation criteria for second-order nonlinear neutral delay difference equations,* Appl. Math. Comput., 142(1): (2003), 99–111.

[13] Zhang Z., Lv X., Yu T.; *Oscillation criteria of second order neutral difference equations,* J. Appl. Math. Comput., **13**(1-2): (2003), 125–138.

Bibliography

Karpuz B., Santra S. S.; *Oscillation theorems for second-order nonlinear delay differential equations of neutral type,* Hacettepe J. Math. Stat., **48**(3): (2019), 633–643.

Pinelas S., Santra S. S.; *Necessary and sufficient condition for oscillation of nonlinear neutral first-order differential equations with several delays,* J. Fixed Point Theory Appl., **20**(1): (2018), Article Id. 27, 1–13.

Pinelas S., Santra S. S.; *Necessary and Sufficient Conditions for Oscillation of Nonlinear First Order Forced Differential Equations with Several Delays of Neutral Type,* Analysis, **39**(3): (2019), 97–105.

Santra S. S.; *Oscillation criteria for nonlinear neutral differential equations of first order with several delays,* Mathematica, **57**(80)(1-2): (2015), 75–89.

Santra S. S.; *Necessary and sufficient condition for oscillation of nonlinear neutral firrst order differential equations with several delays,* Mathematica, **58**(81)(1-2): (2016), 85–94.

Santra S. S.; *Oscillation analysis for nonlinear neutral differential equations of second order with several delays,* Mathematica, **59**(82)(1-2): (2017), 111–123.

Santra S. S.; *Oscillation analysis for nonlinear neutral differential equations of second order with several delays and forcing term,* Mathematica, **61**(84)(1): (2019), 63–78.

Santra S. S., Majumder D., Bhattacherjee R.; *Oscillation Tests: Second-order Neutral Difference Equation,* SSRN (Electronic copy available at: Electronic copy available at: $https://ssrn.com/abstract = 3526477$.)

Tripathy A. K., Santra S. S.; *Necessary and sufficient conditions for oscillation of a class of second order impulsive systems,* Differ. Equ. Dyn. Syst., (published online on 09 May 2018) https://doi.org/10.1007/s12591-018-0425-7

Chapter 6

PDEs Satisfied by the Density Function of Stochastic Integrals

Julia Calatayud, Juan Carlos Cortés and Marc Jornet*
Institut Universitari de Matemàtica Multidisciplinar,
Universitat Politècnica de València,
Valencia, Spain

Abstract

In this chapter, we derive a partial differential equation (PDE) for the density function of stochastic processes of the form $X(t) = \int_{t_0}^{t} Y(s)\,\mathrm{d}s$, $t \geq t_0$, where Y is any stochastic process with values in $\mathbb{R}$. The resulting PDE depends on the conditional expectation of $Y(t)|X(t) = x$, which is in general unknown. However, the PDE is applicable in the case of random differential equation problems, which yields Liouville's equation for the density of the solution. It is also applicable to deduce a PDE analog of the random variable transformation method. On the other hand, when Y is Gaussian, such conditional law $Y(t)|X(t) = x$ is known; therefore explicit PDEs can be obtained and tested on important Gaussian processes (Brownian motion, Brownian bridge, white noise, Ornstein-Uhlenbeck and fractional Brownian motion processes). In particular, when $Y(t) = h(t)W(t)$, where h is a deterministic function and W is a Gaussian white noise process, we find a PDE for the density function of Itô-type integrals, $X(t) = \int_{t_0}^{t} h(s)\,\mathrm{d}B(s)$, where B is a Brownian motion.

---

*Corresponding Author's Email: marjorsa@doctor.upv.es.

Keywords: stochastic integral, probability density function, partial differential equation, Liouville's equation for random differential equations, Gaussian process

AMS Subject Classification: 60H05, 60G15

1. INTRODUCTION

Given a stochastic process $Y(t)$ with values in $\mathbb{R}$, $t \in I$, where I is a real interval whose left endpoint is t_0, one can define a new stochastic process via integration: $X(t) = \int_{t_0}^{t} Y(s)\,\mathrm{d}s$, $t \in [t_0, T]$. Such integral may be properly defined in several ways. The simplest way consists in using the sample paths of Y and their Riemann or real Lebesgue integral. Another possibility is to consider the mean square limit of the Riemann sums, and thus construct the mean square integral [1,2]. Based on the latter sense, integration with respect to a Gaussian white noise process (formal derivative of the Brownian motion) can be performed; the complete mathematical construction of such type of integrals is done via Itô calculus [1,3,4].

Given the probability distribution of the stochastic process Y, the probability law of its integral X is unknown. Only when Y is a Gaussian process, it is known that its integral is Gaussian too [1, Th. 4.6.4]. In this work, we aim at deriving a partial differential equation (PDE) satisfied by the probability density function of $X(t)$, $f_{X(t)}(x)$. The probability density function (or just density function) is characterized by the properties $f_{X(t)} \geq 0$ and $\mathbb{P}[X(t) \in A] = \int_A f_{X(t)}(x)\,\mathrm{d}x$, where A is any Borel set in $\mathbb{R}$. In the general case, we will see that the resulting PDE depends upon the conditional expectation of $Y(t)|X(t) = x$, which is in general unknown. In the case of random differential equations, we will recover Liouville's equation for the density function of the solution [1, Th. 6.2.2]. We will also deduce a PDE version of the random variable transformation method [1, Section 2.4.2], [5, 6]. On the other hand, when Y is a Gaussian process, that conditional law $Y(t)|X(t) = x$ is known; therefore explicit PDEs will be obtained and tested on important Gaussian processes (Brownian motion, Brownian bridge, white noise, Ornstein-Uhlenbeck and fractional Brownian motion processes). When $Y(t) = h(t)W(t)$, where h is a deterministic function and W is a Gaussian white noise process, the integral $X(t)$ becomes an Itô integral; in this case, we will also derive a PDE for the density function of $X(t)$.

2. PDE FOR THE DENSITY FUNCTION OF A GENERAL STOCHASTIC INTEGRAL

Consider a complete probability space $(\Omega, \mathcal{F}, \mathbb{P})$ (triplet formed by the sample space, the σ-algebra of events and the probability measure, respectively) and a stochastic process $Y(t)$ defined on a real interval I with realizations in $\mathbb{R}$. Let t_0 be the left endpoint of I. Consider the integral stochastic process $X(t) = \int_{t_0}^{t} Y(s)\,\mathrm{d}s$, $t \in I$, [1, 2]. Let $f_{X(t)}(x)$ be the probability density function of $X(t)$ at $x \in \mathbb{R}$. The main goal is to derive a PDE for $f_{X(t)}(x)$, in terms of t and x. The reader should notice that, in general, the probability distribution of a stochastic integral is unknown; only when Y is Gaussian, it is known that X is also Gaussian [1, Th. 4.6.4].

Theorem 2.1. *Let $Y(t)$, $t \in I$, be any integrable stochastic process, and let $X(t) = \int_{t_0}^{t} Y(s)\,\mathrm{d}s$, $t \in I$. If $X(t)$ has a density function $f_{X(t)}(x)$ for each $t \in I$, then*

$$\frac{\partial}{\partial t} f_{X(t)}(x) + \frac{\partial}{\partial x}\left\{ f_{X(t)}(x) \mathbb{E}[Y(t)|X(t)=x] \right\} = 0. \tag{1}$$

Here $\mathbb{E}$ is the expectation operator and the vertical bar means conditioning.

Proof. In order to prove (1), let us fix a right endpoint $T > t_0$ of I and let us consider any function $\varphi(x,t) \in C_c^{\infty}(\mathbb{R} \times (t_0, T))$ (smooth function with compact support in $\mathbb{R} \times (t_0, T)$). By Barrow's rule and the relation $\frac{\mathrm{d}}{\mathrm{d}t} X(t) = Y(t)$,

$$0 = \varphi(X(T), T) - \varphi(X(t_0), t_0) = \int_{t_0}^{T} \frac{\partial}{\partial t}\left(\varphi(X(t), t)\right) \mathrm{d}t$$

$$= \int_{t_0}^{T} \left(\frac{\partial \varphi}{\partial t}(X(t), t) + \frac{\partial \varphi}{\partial x}(X(t), t) Y(t) \right) \mathrm{d}t.$$

We apply the expectation operator. On the one hand,

$$\mathbb{E}\left[\frac{\partial \varphi}{\partial t}(X(t), t)\right] = \int_{\mathbb{R}} \frac{\partial \varphi}{\partial t}(x, t) f_{X(t)}(x)\,\mathrm{d}x.$$

On the other hand, using conditional expectations,

$$\mathbb{E}\left[\frac{\partial \varphi}{\partial x}(X(t), t) Y(t)\right] = \int_{\mathbb{R}} \frac{\partial \varphi}{\partial x}(x, t) \mathbb{E}\left[Y(t)|X(t)=x\right] f_{X(t)}(x)\,\mathrm{d}x.$$

Then

$$0 = \int_{t_0}^{T} \int_{\mathbb{R}} \left(\frac{\partial \varphi}{\partial t}(x,t) f_{X(t)}(x) + \frac{\partial \varphi}{\partial x}(x,t) \mathbb{E}\left[Y(t)|X(t) = x\right] f_{X(t)}(x) \right) dx\, dt.$$

Using the integration by parts formula and the compacity of the support of φ,

$$0 = \int_{t_0}^{T} \int_{\mathbb{R}} \left(\frac{\partial}{\partial t} f_{X(t)}(x) + \frac{\partial}{\partial x} \left\{ f_{X(t)}(x) \mathbb{E}[Y(t)|X(t) = x] \right\} \right) \varphi(x,t)\, dx\, dt.$$

By the fundamental lemma of calculus of variations, [7, Lemma 1.1.1, p. 6], (1) follows. □

Remark 2.2. Consider a random differential equation problem

$$\begin{cases} X'(t) = g(X(t), t),\ t \in I, \\ X(t_0) = X_0, \end{cases}$$

where the initial condition X_0 is a random variable and the map $g(x,t)$ is a deterministic function [1,8,9]. If we denote $Y(t) = g(X(t),t)$, then

$$\mathbb{E}[Y(t)|X(t) = x] = \mathbb{E}[g(X(t),t)|X(t) = x] = g(x,t).$$

Using (1), we have rigorously proved that

$$\frac{\partial}{\partial t} f_{X(t)}(x) + \frac{\partial}{\partial x} \left\{ f_{X(t)}(x) g(x,t) \right\} = 0.$$

This expression is known as Liouville's equation. Different proofs of the Liouville's equation have been developed in the literature. The paper [10] presents in the appendix a proof using characteristic functions. This proof is also presented in [1, Th. 6.2.2]. Another proof using dynamical systems appears in [11, Th. 8.4]. Liouville's equation is a particular case of Fokker-Planck's PDE (or forward Kolmogorov equation) for stochastic differential equations of Itô-type, with drift $g(x,t)$ and diffusion coefficient 0.

Although $\mathbb{E}[Y(t)|X(t) = x]$ is in general unknown, let us see some simple examples in which a PDE for $f_{X(t)}(x)$ can be explicitly derived.

Example 2.3. Let $Y(t) = Y$ be an absolutely continuous random variable (i.e. it has a density function f_Y). Then $X(t) = \int_0^t Y(s)\, ds = Yt$, $t > 0$, so

$\mathbb{E}[Y(t)|X(t) = x] = \mathbb{E}[Y|Yt = x] = \mathbb{E}[Y|Y = x/t] = x/t$. Expression (1) becomes

$$\frac{\partial}{\partial t} f_{X(t)}(x) + \frac{1}{t}\frac{\partial}{\partial x}\left\{x f_{X(t)}(x)\right\} = 0.$$

Notice that this equation is clear, just by differentiating the relation $f_{X(t)}(x) = f_Y(\frac{x}{t})\frac{1}{t}$ (random variable transformation method, [1, Section 2.4.2]).

Example 2.4. Let $X(t) = e^{At}$, $t > 0$, where A is any absolutely continuous random variable. Let $Y(t) = X'(t) = AX(t)$. The conditional expectation of $Y(t)$ given that $X(t) = x$, $x > 0$, is

$$\mathbb{E}[Y(t)|X(t) = x] = x\mathbb{E}[A|X(t) = x] = \frac{x \log x}{t}.$$

Hence

$$\frac{\partial}{\partial t} f_{X(t)}(x) + \frac{1}{t}\frac{\partial}{\partial x}\left\{x \log x\, f_{X(t)}(x)\right\} = 0.$$

This PDE can also be obtained by differentiating the identity $f_{X(t)}(x) = f_A(\frac{1}{t}\log x)\frac{1}{xt}$ (this identity comes from the random variable transformation method, [1, Section 2.4.2]).

Example 2.5. Let $X(t) = \sin(At)$, $0 < t < \pi/2$, A absolutely continuous random variable in $(0, 1)$. Let $Y(t) = X'(t) = A\cos(At)$. Notice that $X(t) = x \in (0, 1)$ if and only if $A = \arcsin(x)/t$. In such a case, $Y(t) = \arcsin(x)/t \cdot \sqrt{1 - x^2}$. Hence

$$\frac{\partial}{\partial t} f_{X(t)}(x) + \frac{1}{t}\frac{\partial}{\partial x}\left\{\arcsin(x)\sqrt{1 - x^2}\, f_{X(t)}(x)\right\} = 0.$$

This PDE is checked to be true by differentiating $f_{X(t)}(x) = f_A(\frac{\arcsin x}{t})\frac{1}{t\sqrt{1-x^2}}$ (this identity is a consequence of the random variable transformation technique, [1, Section 2.4.2]).

These examples can be encompassed under the following general result. It may be viewed as a PDE analog of the random variable transformation method.

Corollary 2.6. *Let $X(t) = g(A, t)$ be a deterministic transformation of an absolutely continuous random variable A. Suppose that, for each t, the function $g(\cdot, t)$ is invertible with inverse $g^{-1}(\cdot, t)$. Then*

$$\frac{\partial}{\partial t} f_{X(t)}(x) + \frac{\partial}{\partial x}\left\{\partial_2 g(a,t)|_{a=g^{-1}(x,t)} f_{X(t)}(x)\right\} = 0.$$

Proof. Notice that $X(t) = x$ if and only if $A = g^{-1}(x,t)$. Given $Y(t) = X'(t) = \partial_2 g(A,t)$, we deduce that $\mathbb{E}[Y(t)|X(t) = x] = \partial_2 g(a,t)|_{a=g^{-1}(x,t)}$. The statement thus follows from Theorem 2.1. □

3. PDE FOR THE DENSITY FUNCTION OF A GAUSSIAN STOCHASTIC INTEGRAL WITH RESPECT TO THE LEBESGUE MEASURE

In the PDE (1), the conditional law $Y(t)|X(t) = x$ is in general unknown. However, in the case of a Gaussian process Y, such conditional distribution is known and more explicit formulas of (1) can be obtained.

For the sake of completeness, we recall that a stochastic process Y is Gaussian if each random vector $(Y(t_1), \ldots, Y(t_k))$, $t_1, \ldots, t_k \in I$, $k \geq 1$, is Gaussian distributed; that is, every linear combination $\sum_{i=1}^{k} a_i Y(t_i)$, $a_1, \ldots, a_k \in \mathbb{R}$, is one-dimensional Gaussian. Equivalently, its characteristic function is given by

$$\phi_{(Y(t_1),\ldots,Y(t_k))}(u) = \mathbb{E}\left[e^{i\sum_{j=1}^{k} t_j u_j}\right] = e^{i\mu^\top u - \frac{1}{2} u^\top \Lambda u},$$

where $u = (u_1, \ldots, u_k)^\top \in \mathbb{R}^k$, $\top$ denotes the transpose operator, μ is the mean vector, and Λ is the covariance matrix, [1, p. 66].

Theorem 3.1. *Let $Y(t)$, $t \in I$, be any Gaussian integrable stochastic process, and let $X(t) = \int_{t_0}^{t} Y(s) \, \mathrm{d}s$, $t \in I$. Then,*

$$\frac{\partial}{\partial t} f_{X(t)}(x) + \frac{\partial}{\partial x} \left\{ f_{X(t)}(x) \left(\mathbb{E}[Y(t)] + \frac{\mathrm{Cov}[Y(t), X(t)]}{\mathbb{V}[X(t)]} (x - \mathbb{E}[X(t)]) \right) \right\} = 0, \quad (2)$$

where

$$\mathbb{V}[X(t)] = \int_{t_0}^{t} \int_{t_0}^{t} \mathrm{Cov}[Y(s), Y(\tau)] \, \mathrm{d}s \, \mathrm{d}\tau,$$

$$\mathrm{Cov}[Y(t), X(t)] = \int_{t_0}^{t} \mathrm{Cov}[Y(t), Y(s)] \, \mathrm{d}s.$$

Here $\mathbb{E}$, $\mathbb{V}$ and Cov are the expectation, variance and covariance operators, respectively.

Proof. By Theorem 2.1, we need to prove the relation

$$\mathbb{E}[Y(t)|X(t) = x] = \mathbb{E}[Y(t)] + \frac{\mathbb{C}\mathrm{ov}[Y(t), X(t)]}{\mathbb{V}[X(t)]}(x - \mathbb{E}[X(t)]). \quad (3)$$

Let us first see that the random vector $(Y(t), X(t))$ is Gaussian, for each $t \in I$. We can write $X(t)$ as a limit of Riemann sums:

$$X(t) = \int_{t_0}^{t} Y(\tau)\,\mathrm{d}\tau = \lim_{n \to \infty} \sum_{i=1}^{r_n} Y(t_i^n)(t_{i+1}^n - t_i^n),$$

where $t_0 = t_1^n < t_2^n < \ldots < t_{r_n}^n < t_{r_n+1}^n = t$ is a partition of $[t_0, t]$, for each $n \geq 1$, and the limit is almost surely (for sample path integrals) or mean square. Then,

$$(Y(t), X(t)) = \lim_{n \to \infty} (Y(t), \sum_{i=1}^{r_n} Y(t_i^n)(t_{i+1}^n - t_i^n)). \quad (4)$$

Notice that

$$\begin{pmatrix} Y(t) \\ \sum_{i=1}^{r_n} Y(t_i^n)(t_{i+1}^n - t_i^n) \end{pmatrix} = \begin{pmatrix} 1 & 0 & \cdots & 0 \\ 0 & t_2^n - t_1^n & \cdots & t_{r_n+1}^n - t_{r_n}^n \end{pmatrix} \begin{pmatrix} Y(t) \\ Y(t_1^n) \\ \vdots \\ Y(t_{r_n}^n) \end{pmatrix},$$

and since $(Y(t), Y(t_1^n), \ldots, Y(t_{r_n}^n))$ is a Gaussian random vector, we conclude that

$$(Y(t), \sum_{i=1}^{r_n} Y(t_i^n)(t_{i+1}^n - t_i^n))$$

is also Gaussian. By (4), the random vector $(Y(t), X(t))$ must be Gaussian, as wanted.

Finally, by [13, Example 4.51], $Y(t)|X(t) = x$ has Gaussian law and (3) holds. $\square$

Let us see some examples of applicability of Theorem 3.1, for different Gaussian processes $Y(t)$.

Example 3.2. Let $Y(t) = B(t)$ be a standard Brownian motion on $I = [0, \infty)$ [13, Ch. 5]. It is defined as a Gaussian process with mean zero and covariance function $\mathbb{C}\mathrm{ov}[B(t), B(s)] = \min\{t, s\}$. From this definition, it is easily deduced that $B(0) = 0$ almost surely, B has independent and stationary increments, and it is self-similar. The Brownian motion has a modification which is continuous almost surely, but it is nowhere differentiable. It is known that $X(t) = \int_0^t Y(s)\,\mathrm{d}s \sim \text{Normal}(0, \frac{t^3}{3})$ [2, Example 3.20]. On the other hand,

$$\mathrm{Cov}[Y(t), X(t)] = \int_0^t \min\{t, s\}\,\mathrm{d}s = \int_0^t s\,\mathrm{d}s = \frac{t^2}{2}.$$

As a consequence, the PDE (2) is

$$\frac{\partial}{\partial t} f_{X(t)}(x) + \frac{3}{2t} \frac{\partial}{\partial x}\left(x f_{X(t)}(x)\right) = 0.$$

Example 3.3. Consider $Y(t)$ as a standard Brownian bridge on $I = [0, 1]$, [13, p. 193]. It is a mean-zero Gaussian process with covariance $\mathrm{Cov}[Y(t), Y(s)] = \min\{t, s\} - ts$. Let $X(t) = \int_0^t Y(s)\,\mathrm{d}s$. From these properties and by simple computations, one obtains

$$\mathbb{V}[X(t)] = \int_0^t \int_0^t (\min\{s, \tau\} - s\tau)\,\mathrm{d}s\,\mathrm{d}\tau = \frac{t^3}{3} - \frac{t^4}{4}$$

and

$$\mathrm{Cov}[Y(t), X(t)] = \int_0^t (\min\{t, s\} - ts)\,\mathrm{d}s = \frac{(1-t)t^2}{2}.$$

The PDE (2) thus becomes

$$\frac{\partial}{\partial t} f_{X(t)}(x) + \frac{6(t-1)}{t(3t-4)} \frac{\partial}{\partial x}\left(x f_{X(t)}(x)\right) = 0.$$

Example 3.4. Let $Y(t) = W(t)$ be a Gaussian white noise process. Its formal properties are $\mathbb{E}[Y(t)] = 0$ and $\mathrm{Cov}[Y(t), Y(s)] = \delta(t-s)$ [13, p. 196], where δ is the Dirac delta function. It may be viewed as the formal derivative of the Brownian motion $B(t)$. That is, $X(t) = \int_0^t Y(s)\,\mathrm{d}s$ is a standard Brownian motion on $I = [0, \infty)$. We have $\mathbb{V}[X(t)] = t$ and $\mathrm{Cov}[Y(t), X(t)] = \int_0^t \delta(t-s)\,\mathrm{d}s = 1/2$, so that (2) becomes

$$\frac{\partial}{\partial t} f_{X(t)}(x) + \frac{1}{2t} \frac{\partial}{\partial x}\left(x f_{X(t)}(x)\right) = 0.$$

Example 3.5. Given a Brownian motion $B(t)$, let $Y(t) = B(e^t)/e^{t/2}$. This process is known as Ornstein-Uhlenbeck process [14, Example 3.9]. It is Gaussian, with mean zero and covariance function $\mathbb{C}\text{ov}[Y(t), Y(s)] = e^{-\frac{1}{2}|t-s|}$. Such processes whose covariance function depends on t and s only through $|t-s|$ are called stationary. Let $X(t) = \int_0^t Y(s)\,\mathrm{d}s$. Simple computations yield

$$\mathbb{V}[X(t)] = \int_0^t \int_0^t e^{-\frac{1}{2}|s-\tau|}\,\mathrm{d}s\,\mathrm{d}\tau = 8(e^{-t/2} - 1) + 4t$$

and

$$\mathbb{C}\text{ov}[Y(t), X(t)] = \int_0^t e^{-\frac{1}{2}|t-s|}\,\mathrm{d}s = 2(1 - e^{-t/2}).$$

Then (2) becomes

$$\frac{\partial}{\partial t} f_{X(t)}(x) + \frac{1 - e^{-t/2}}{4(e^{-t/2} - 1) + 2t} \frac{\partial}{\partial x}\left(x f_{X(t)}(x)\right) = 0.$$

Example 3.6. Let $Y(t) = B^H(t)$ be a fractional Brownian motion, [13, p. 196], with Hurst parameter $H \in (0, 1)$. It is defined as a Gaussian process with zero-mean and covariance function $\mathbb{C}\text{ov}[Y(t), Y(s)] = \frac{1}{2}(|t|^{2H} + |s|^{2H} - |t-s|^{2H})$. For $H = 1/2$ we recover the Brownian motion from Example 3.2. Let $X(t) = \int_0^t Y(s)\,\mathrm{d}s$. We have

$$\mathbb{V}[X(t)] = \int_0^t \int_0^t \frac{1}{2}(|s|^{2H} + |\tau|^{2H} - |s-\tau|^{2H})\,\mathrm{d}s\,\mathrm{d}\tau = \frac{t^{2+2H}}{2+2H}$$

and

$$\mathbb{C}\text{ov}[Y(t), X(t)] = \int_0^t \frac{1}{2}(|t|^{2H} + |s|^{2H} - |t-s|^{2H})\,\mathrm{d}s = \frac{t^{1+2H}}{2}.$$

The PDE (2) is

$$\frac{\partial}{\partial t} f_{X(t)}(x) + \frac{1+H}{t} \frac{\partial}{\partial x}\left(x f_{X(t)}(x)\right) = 0.$$

Remark 3.7. Formula (3) shows an interesting geometric property. Notice that the right-hand side of (3) is a straight line in the variable x, for each t. If we want to approximate $Y(t) \approx aX(t) + b$, where a and b are deterministic, the values of a and b that minimize $\mathbb{E}[(Y(t) - aX(t) - b)^2]$ are

$$a = \frac{\mathbb{C}\text{ov}[X(t), Y(t)]}{\mathbb{V}[X(t)]}, \quad b = \mathbb{E}[Y(t)] - a\mathbb{E}[X(t)].$$

On the other hand, the left-hand side of (3) is the best deterministic function $h(x)$ for which $Y(t) \approx h(X(t))$, in the sense of minimizing $\mathbb{E}[\{Y(t) - h(X(t))\}^2] = \int_{\mathbb{R}} \mathbb{E}[(Y(t) - h(x))^2 | X(t) = x] f_{X(t)}(x) \, dx$. Such function is given by $h(x) = \mathbb{E}[Y(t)|X(t) = x]$. Hence, in the setting of Theorem 3.1, the best approximation $h(x)$ coincides with the straight line $ax + b$.

Corollary 3.8. *Let $Y(t)$, $t \in I$, be any Gaussian integrable stochastic process, and let $X(t) = \int_{t_0}^{t} Y(s) \, ds$, $t \in I$. Then the following convection-diffusion equation holds:*

$$\frac{\partial}{\partial t} f_{X(t)}(x) + \mathbb{E}[Y(t)] \frac{\partial}{\partial x} f_{X(t)}(x) = \mathrm{Cov}[Y(t), X(t)] \frac{\partial^2}{\partial x^2} f_{X(t)}(x). \quad (5)$$

Proof. By differentiating the general density function of a Normal distribution directly, it is easy to check that

$$\mathbb{V}[X(t)] \frac{\partial^2}{\partial x^2} f_{X(t)}(x) + \frac{\partial}{\partial x} \{f_{X(t)}(x) (x - \mathbb{E}[X(t)])\} = 0.$$

Combining this equation and (2), we arrive at the required PDE. □

Remark 3.9. *In reference [12], the authors deduced (5) alternatively by differentiating the density function of a Normally distributed integral directly.*

Example 3.10. From Examples 3.2–3.6 and Corollary 3.8, the following diffusion PDEs are deduced when $Y(t)$ is a Brownian motion, Brownian bridge, white noise, Ornstein-Uhlenbeck and fractional Brownian motion process, respectively, and $X(t) = \int_0^t Y(s) \, ds$:

$$\frac{\partial}{\partial t} f_{X(t)}(x) = \frac{t^2}{2} \frac{\partial^2}{\partial x^2} f_{X(t)}(x),$$

$$\frac{\partial}{\partial t} f_{X(t)}(x) = \frac{(1-t)t^2}{2} \frac{\partial^2}{\partial x^2} f_{X(t)}(x),$$

$$\frac{\partial}{\partial t} f_{X(t)}(x) = \frac{1}{2} \frac{\partial^2}{\partial x^2} f_{X(t)}(x),$$

$$\frac{\partial}{\partial t} f_{X(t)}(x) = 2(1 - e^{-t/2}) \frac{\partial^2}{\partial x^2} f_{X(t)}(x)$$

and

$$\frac{\partial}{\partial t} f_{X(t)}(x) = \frac{1}{2} t^{1+2H} \frac{\partial^2}{\partial x^2} f_{X(t)}(x).$$

4. PDE FOR THE DENSITY FUNCTION OF A GAUSSIAN STOCHASTIC INTEGRAL WITH RESPECT TO THE BROWNIAN MOTION

In this section, we consider the stochastic process $Y(t) = h(t)W(t)$, where $h(t)$ is any square integrable deterministic function and $W(t)$ is a Gaussian white noise process, with the following formal properties: $\mathbb{E}[Y(t)] = 0$ and $\mathrm{Cov}[Y(t), Y(s)] = \delta(t-s)$ [13, p. 196], where δ is the Dirac delta function. The integral $X(t) = \int_{t_0}^{t} Y(s)\,\mathrm{d}s$ is formalized via the concept of Itô integral [1,3,4], and is usually denoted as $X(t) = \int_{t_0}^{t} h(s)\,\mathrm{d}B(s)$, where B is a Brownian motion [13, Ch. 5]. Brownian motion is defined as a mean-zero Gaussian process with covariance $\mathrm{Cov}[Y(t), Y(s)] = \min\{t, s\}$, whose trajectories are continuous but nowhere differentiable. The white noise process may be viewed as the formal derivative of the Brownian motion. The law of $X(t)$ is known: Gaussian with mean zero and variance $\|h\|_{2,t}^2 = \int_{t_0}^{t} h(s)^2\,\mathrm{d}s$ (Itô isometry).

We can apply the theory from the previous sections to derive a PDE for the density function $f_{X(t)}(x)$.

Theorem 4.1. *Let $X(t) = \int_{t_0}^{t} h(s)\,\mathrm{d}B(s)$, $t \in I$, where h is any square integrable deterministic function. Then*

$$\|h\|_{2,t}^2 \frac{\partial}{\partial t} f_{X(t)}(x) + \frac{1}{2}h(t)^2 \frac{\partial}{\partial x}\left(x f_{X(t)}(x)\right) = 0.$$

Proof. The PDE comes from Theorem 3.1, taking into account that

$$\mathbb{E}[Y(t)] = \mathbb{E}[X(t)] = 0, \quad \mathbb{V}[X(t)] = \|h\|_{2,t}^2$$

and

$$\mathrm{Cov}[Y(t), X(t)] = \int_{t_0}^{t} h(t)h(s)\mathrm{Cov}[W(t), W(s)]\,\mathrm{d}s$$

$$= h(t)\int_{t_0}^{t} h(s)\delta(t-s)\,\mathrm{d}s = \frac{1}{2}h(t)^2,$$

and substituting into (2). □

Acknowledgments

This work has been supported by the Spanish Ministerio de Economía y Competitividad grant MTM2017–89664–P. The author Marc Jornet acknowledges the doctorate scholarship granted by Programa de Ayudas de Investigación y Desarrollo (PAID), Universitat Politècnica de València.

Conflict of Interest Statement

The authors declare that there is no conflict of interests regarding the publication of this article.

Conclusion

In this chapter we obtained a PDE for the density function of stochastic integrals. As it depends on a conditional law which is in general unknown, this PDE has limited applicability in practice except for certain examples. Nonetheless, Liouville's equation for random differential equations and a PDE version of the random variable transformation method are derived from our PDE expression. In the particular case of Gaussian processes, the PDE has an explicit expression that can be evaluated in important cases, namely when the integrand is a Brownian motion, Brownian bridge, white noise, Ornstein-Uhlenbeck or fractional Brownian motion process. The case of the white noise process gives rise to the Itô integral. More research could be conducted in the future to find new PDEs for density functions.

References

[1] Soong, T. T. (1973). *Random Differential Equations in Science and Engineering*. New York: Academic Press.

[2] Villafuerte, L., Braumann, C. A., Cortés, J.-C. and Jódar, L. (2010). Random differential operational calculus: Theory and applications. *Computers and Mathematics with Applications*, 59: 115–125.

[3] Øksendal, B. (2013). *Stochastic Differential Equations: An Introduction with Applications*. Berlin Heidelberg: Springer Science & Business Media.

[4] Evans, L. C. (2012). *An Introduction to Stochastic Differential Equations*. American Mathematical Society.

[5] Dorini, F. A., Cecconello, M. S. and Dorini, L. B. (2016). On the logistic equation subject to uncertainties in the environmental carrying capacity and initial population density. *Communications in Nonlinear Science and Numerical Simulation*, 33: 160–173.

[6] Cortés, J.-C., Navarro-Quiles, A., Romero, J. V. and Roselló, M. D. (2018). Computing the probability density function of non-autonomous first-order linear homogeneous differential equations with uncertainty. *Journal of Computational and Applied Mathematics*, 337: 190–208.

[7] Jost, J. and Li-Jost, X. (1998). *Calculus of Variations*. New York: Cambridge University Press.

[8] Neckel, T. and Rupp, F. (2013). *Random Differential Equations in Scientific Computing*. London: Walter de Gruyter.

[9] Strand, J. L. (1970). Random ordinary differential equations. *Journal of Differential Equations*, 7 (3): 538–553.

[10] Kozin, F. (1961). On the probability densities of the output of some random systems. *Journal of Applied Mechanics*, 28: 161–165.

[11] Saaty, T. L. (1981). *Modern Nonlinear Equations*. New York: Dover.

[12] Dorini, F. A. and Cunha, M. C. C. (2011). On the linear advection equation subject to random velocity fields. *Mathematics and Computers in Simulation*, 82: 679–690.

[13] Lord, G. J., Powell, C. E. and Shardlow, T. (2014). *An Introduction to Computational Stochastic PDEs*. New York: Cambridge University Press.

[14] Khoshnevisan, D. (2009). A Primer on Stochastic Partial Differential Equations. In *A Minicourse on Stochastic Partial Differential Equations* (pp. 1–38). Berlin Heidelberg: Springer.

INDEX

A

abstract Volterra integro-differential inclusions, vii, 2

B

Banach spaces, vii, 5, 25
biotechnology, 50
bounded linear operators, 31
Brownian motion, viii, 107, 108, 114, 115, 116, 117, 118

C

calculus, 2, 4, 108, 118
continuous random variable, 110, 111
cosine family, 29, 30, 31, 32, 47

D

delay, viii, 49, 58, 61, 62, 81, 82, 83, 84, 85, 103, 104
derivatives, 4, 42
difference equation, viii, 59, 85, 86, 103, 104
differential equations, vii, 4, 23, 26, 49, 58, 59, 61, 62, 79, 81, 82, 83, 84, 103, 104, 110, 118, 119
distribution, 82, 112

E

engineering, 4, 50, 62, 86
evolution, 25, 26

F

families, 30, 48

G

Gaussian process, 108, 112, 113

I

impulses, 58, 59
impulsive, vii, 49, 59, 105
impulsive differential equations, 49, 59
inequality, 20, 40, 55, 99, 100, 101

K

Knaster-Tarski fixed point theorem, viii

L

Lebesgue spaces with variable exponents, 1, 2, 3, 7, 21, 25

M

mapping, 4, 16, 18, 20, 21, 22, 24, 31, 34, 74, 100
matrix, 112
metallurgy, 50
mild and strict solutions, 29
modelling, 62
models, 62
motivation, 102

N

natural science, 62
neutral, vii, viii, 49, 50, 51, 53, 55, 57, 58, 59, 61, 62, 64, 81, 82, 83, 85, 86, 87, 95, 103, 104
neutral differential equations, vii, 59, 61, 82, 83, 104
nonoscillation, 49, 61, 62, 85

O

OSC, 49, 54, 64, 70, 85, 87, 95
oscillation, vii, viii, 49, 50, 51, 53, 55, 57, 58, 59, 61, 62, 63, 64, 81, 82, 83, 84, 85, 86, 103, 104, 105

P

PAA, 8, 17
partial differential equation, viii, 86, 119
partition, 113
PDEs, viii, 116
pharmacokinetics, 50
probability, 108, 109, 119
probability density function, 108, 109, 119

probability distribution, 108
proposition, 14, 15, 20, 21

R

radiation, 86
real numbers, 17, 50, 51, 87, 95
robotics, 50

S

second order partial differential, 29
solution, viii, 24, 32, 33, 34, 39, 40, 42, 44, 47, 50, 51, 52, 54, 56, 62, 63, 64, 65, 71, 73, 74, 76, 80, 81, 83, 86, 87, 88, 89, 90, 92, 93, 95, 96, 98, 100, 101, 102, 104, 107
statistics, 86
Stepanov, vii, 1, 2, 3, 7, 8, 9, 10, 11, 12, 15, 16, 18, 20, 22, 25, 26
Stepanov almost automorphy with variable exponents, 1
stochastic integral, 108, 109, 118
stochastic processes, viii

T

technology, 50, 86
transformation, viii, 62, 102, 107, 108, 111, 118
translation, 2, 10, 11, 12, 20
transmission, 62

V

vector, 8, 9, 10, 17, 18, 112, 113
velocity, 119